D0138572

PREFACE

The purpose of this text is to provide a basic knowledge of avionics. The word was coined as a contraction of *avi*ation electr*onics*. In its most encompassing meaning, it includes electronics, electrical power, and communication and navigation equipment. The term originally applied to airborne equipment, but it often includes ground-based aviation-related equipment and aerospace systems.

Avionics systems can be as simple as discrete units operating independently, or as complex as integrated systems where many units are essentially "talking" to one another. Avionics generally can be subdivided into three areas: communications, power, and navigation. This text focuses primarily on navigation, with some exceptions—notably the Flight Control chapter. In order to design, maintain, repair, and service aviation electronics it is critical to understand how these systems operate. In fact, the whole aim of the technician is to keep all equipment in good working order so that the pilot can accurately and safely fly the aircraft.

The authors' intention is to provide a basic understanding of avionics that addresses the needs of students at college and university levels, and perhaps to enhance or reinforce working technicians' knowledge.

The primary audience for this book are those interested in avionics, aircraft maintenance, and aeronautical design. Pilot perspective is also provided as supplemental to the main goal.

This is an instructional text for training and as such provides a simple overview of electronic systems; it is not meant to be an in-depth study of avionics engineering. In all cases of actual maintenance, the manufacturers' manuals must be followed.

For ease of reading, the text contains many outline-style lists. We have deliberately attempted to avoid long-winded highly technical explanations and to "keep it simple" throughout.

The glossary includes abbreviations, acronyms, and jargon—the esoteric language of the avionics industry. The appendix includes supplemental information.

Each chapter follows the same general outline: introduction, objectives, basic concept, theory, pilot perspective, basic electronics, more detailed electronics, and questions. Solutions to the questions are provided at the end of the text.

For readers who wish to examine the electronics in more detail, we have tried to provide more of the technical details. Some systems are so complex that the only possible way to describe them is with the actual functional electronics.

We can expect a number of avionics systems to be phased out primarily due to the microprocessor and satellites. This of course means that we can expect to see some exciting new things happen. We have described the traditional systems, and even in some cases the obsolete ones, to put some perspective on the changes that are occurring.

ACKNOWLEDGMENTS

To all of the colleagues and individuals who made significant contributions to *Introduction to Avionics*—Thank you. More specifically, we thank E. A. Seddon, C. G. McDonald, and R. S. Brown for technical support, and Jennifer Auger for photographic contribution.

CONTENTS

INTRODUCTION

A very large portion of the electronics in an aircraft is used to determine the position of the aircraft relative to a fixed point on the ground. This can be achieved in several ways. Once the position of the aircraft is known, the pilot can then take steps to navigate by using rate and time relationships.

Radio signals can be manipulated to measure either distance or bearing. Several combinations of these parameters can be combined to determine the aircraft's position.

For example, if both the distance and the bearing to one ground station are known, the pilot can calculate the aircraft's position from this information. This is called Rho Theta navigation, Rho standing for distance and Theta for bearing. If the distance to two ground stations is known, this will provide the pilot with two possible positions. This ambiguity can be eliminated by using the distance information to a third ground station. This is called Rho Rho Rho navigation. Aircraft position can also be determined by measuring the bearing to two or more ground stations. This is called Theta Theta navigation. Two of the methods appear in Figure i.1.

Figure i.1
Two Methods of Navigation

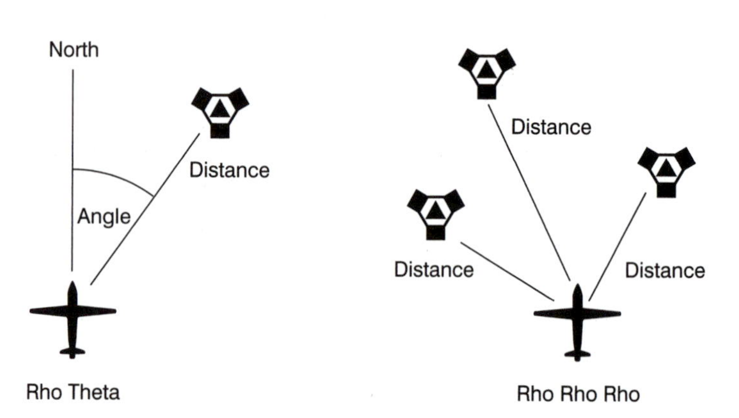

Any combination of these three navigation methods is the basis for most avionics navigation systems. The exceptions are the Doppler Navigation System and the Inertial Navigation System. Doppler navigation utilizes onboard radar, ground echoes, and a variety of computers to obtain the navigation solutions. Inertial navigation is little more than an expansion of the simplest form of navigation called dead reckoning, which is the process of establishing a navigation fix by knowing the aircraft's starting point, direction, speed, and flying time.

CHAPTER 1

Instrument Landing System

ILS

Introduction

This chapter deals with the basic principles of the Instrument Landing System (ILS).

Objectives

State the purpose of ILS

Describe the radiation pattern of the localizer (LOC)

Describe the radiation pattern of the glide slope (G/S)

Describe modulation components for localizer and glide slope signals

Describe the radiation pattern of marker beacon (MKR) signals

Describe the modulation component of the marker signal

Describe aural and illumination signals for the marker

Identify major blocks on a block diagram of glide slope and localizer

Identify major blocks on a block diagram of marker beacon

Explain the Categories of ILS equipment

Basic Concept

The Instrument Landing System (ILS) assists a pilot in positioning the aircraft for landing under low visibility conditions such as fog, snow, low clouds, rain, and darkness.

ILS THEORY

The **Instrument Landing System (ILS)** assists a pilot in positioning the aircraft for landing under low visibility conditions such as fog, snow, low clouds, rain, and darkness. ILS is a VHF/UHF radio navigation aid that provides two radio beams which can be used as an ideal flight path. This flight path extends to about 40 mi from the end of the runway. Two transmitters are located at the runway. One transmitter and its associated antenna array provide for azimuth information. This one is the localizer. The other transmitter and its associated antenna array provide for elevation information. This one is called the glide slope (G/S).

Both localizer and glide slope radio transmitters send two electromagnetic energy patterns that overlap one another. The narrow area of overlap defines the ideal flight path. The localizer and the G/S both obtain position information by comparing the relative strength of two audio tones.

The ILS provides information about

- azimuth reference
- approximate range
- elevation reference

The frequency range of the localizer is from 108 MHz to 112 MHz, more specifically from 108.1 to 111.95 MHz, using all of the odd frequencies at 50 kHz channel spacing. There are 40 channels. The localizer transmitter operates at about the 100 W power level.

Localizer and VOR (explained in Chapter 5) usually share the same receiver, but internally use different circuitry to convert the information, as either VOR (directional information) or localizer (landing information).

The **localizer** radiates two lobes. One is modulated at 90 Hz and the other is modulated at 150 Hz. The two lobes overlap at equal strength along the centerline of the runway and create a zone of overlap in the center, where both signals have the same amplitude. This provides the vertical plane or the azimuth reference for aircraft approaching the runway. (See Figure 1.1.)

The **glide slope** (Figure 1.2) radiates two lobes. Like the localizer, one is modulated at 90 Hz and the other at 150 Hz. The glide slope provides a glide path that is typically 3° to the horizontal. The glide slope, which uses a much higher frequency than the localizer, operates in the frequency range from 328 MHz to 336 MHz. The combination of localizer and glide slope is referred to as the "front course approach."

It should be noted that the glide slope and localizer frequencies are *paired.* For any given localizer frequency there is a defined glide slope frequency. The navigation tuning is done on the NAV receiver in the cockpit,

Figure 1.1
Localizer Front Course

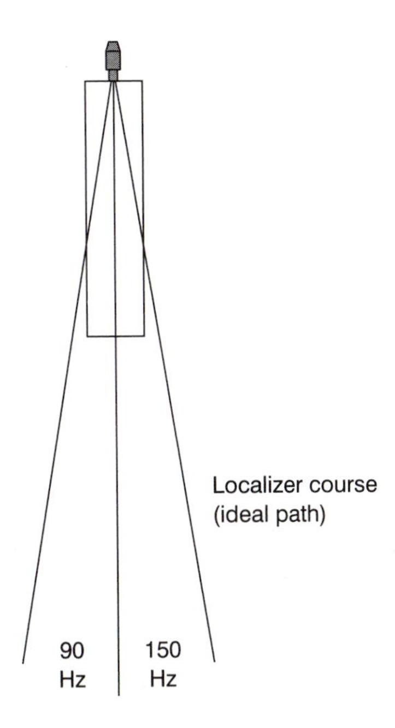

and is tuned to the VHF frequency; the UHF frequency is then automatically tuned, but the pilot does not see this.

There exists a localizer radiation pattern from the back side of the active localizer system. This radiation pattern is similar to the front course with some notable differences, as shown in Figure 1.3. The glide slope information is *not* available on the back course approach. The 90 Hz and 150 Hz modulated tones are on the same physical side of the runway as the front course. On a back course approach the aircraft is flying in the opposite direction as compared to a front course approach. This means that from the point of view of the aircraft the 90 Hz and 150 Hz tones appear to be on the wrong side. The pilot must be aware of this important difference between flying the front course and flying the back course. In the front course approach, the procedure to correct for positional error is to "fly to the needle." In the back course approach, the pilot must do the opposite, and "fly away from the needle."

Figure 1.2
Ideal Glide Slope

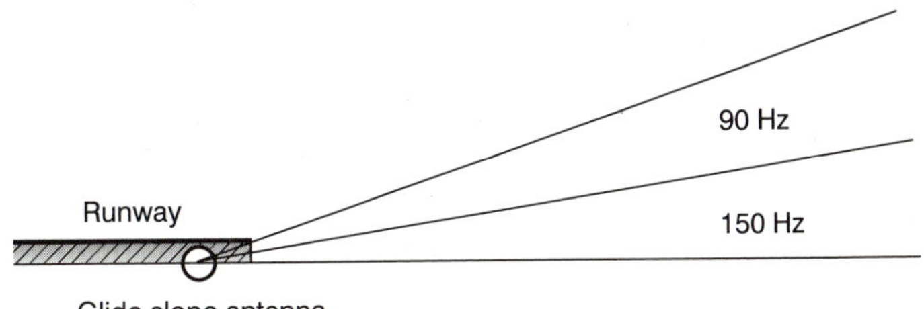

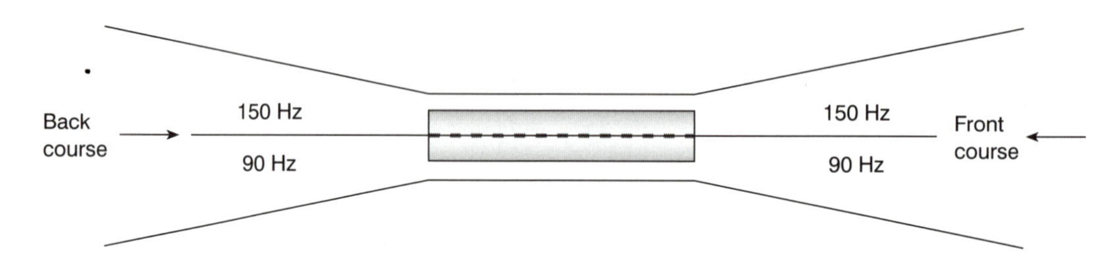

Figure 1.3
Satellite View of Inbound Front and Back Courses

Marker Beacons

The distance information is given to the approaching aircraft from ground antennas about 6 mi from the end of the runway. This transmitter is called the **outer marker** beacon and transmits a vertical cone (elliptical) signal. The other distance marker, the **middle marker** beacon, is about 3500 ft from the end of the runway, and also transmits a vertical cone signal. (See Figures 1.4 and 1.5.)

The marker beacons are low-power transmitters of about 5 W and transmit a vertical cone to provide for an exact reference location. They operate at a 90% modulation level and the receivers are usually of low sensitivity. On aeronautical charts these marker beacons are drawn as eyeball shapes. Marker beacons use a transmitter signal at a carrier frequency of 75 MHz, with one of three modulating frequencies:

- The outer marker beacon is amplitude modulated at 400 Hz and keyed to emit a series of Morse code dashes. An associated blue or purple lamp flashes in the cockpit.

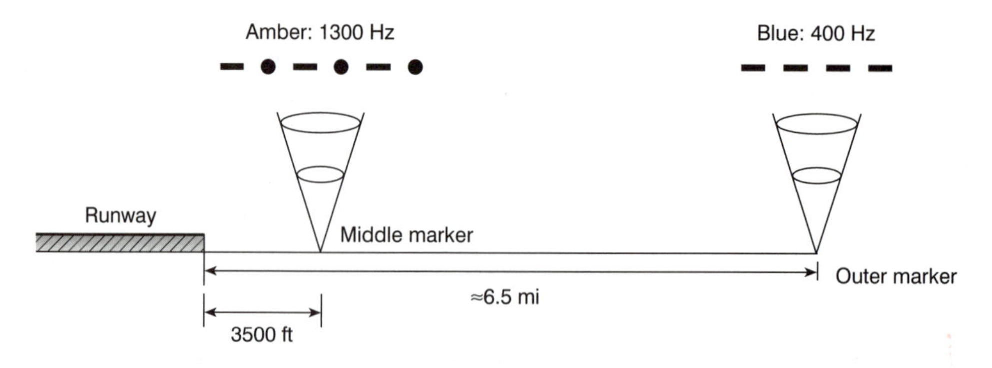

Figure 1.4
Marker Beacons

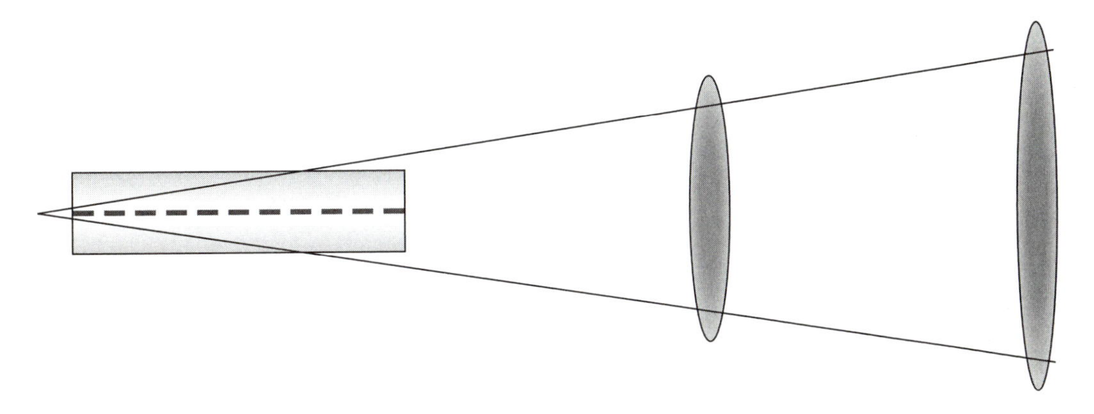

Figure 1.5
Top View of the Runway and Marker Beacon Positions

- The middle marker beacon is amplitude modulated at 1300 Hz and keyed to emit a series of Morse code dots and dashes. An associated amber lamp flashes in the cockpit.
- The fan or airways marker beacon is amplitude modulated at 3000 Hz and keyed to emit a series of Morse code dots. An associated white lamp flashes in the cockpit.

If the aircraft on the localizer is at circuit altitude, the aircraft usually would intercept the glide slope and start the descent at the outer marker.

The use of the marker beacon system on final approach has been decommissioned. The pilot's decision to begin the descent is now based on a combination of glide slope intercepts and Non Directional Radio Beacons (NDBs) station passage (see Chapter 6, Automatic Direction Finder, ADF). However, fan markers (FM) still exist on airways.

Figure 1.6 shows a marker panel.

PILOT PERSPECTIVE

The Canadian Air Regulations prescribe two types of Flight Rules:

- **Visual Flight Rules (VFR)** and
- **Instrument Flight Rules (IFR)**

A pilot is said to be "flying VFR" when the ground is in sight at all times. A pilot "flying IFR" must rely entirely on the aircraft instruments for

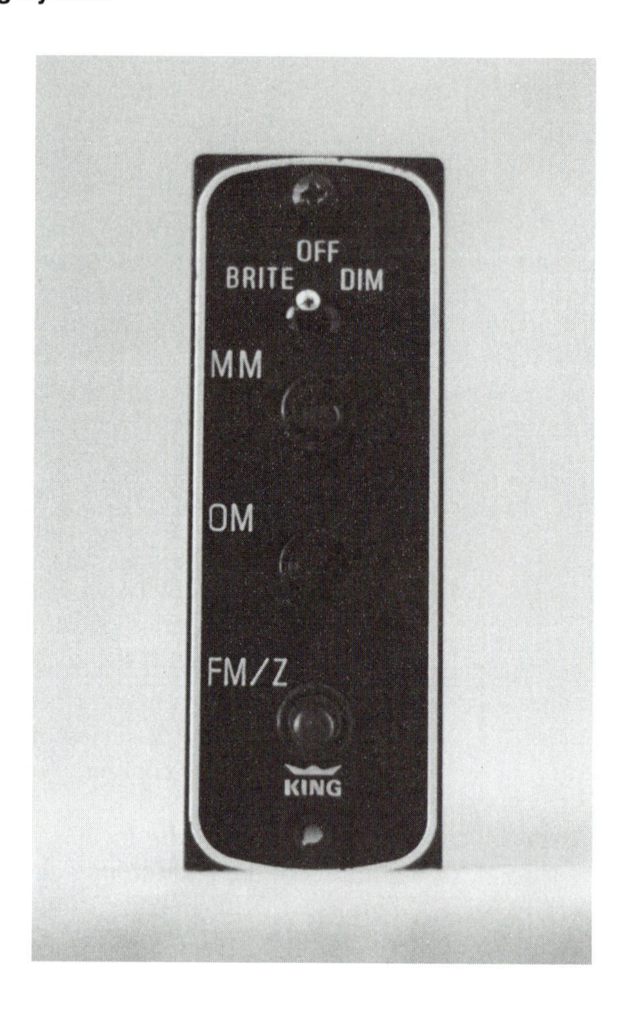

Figure 1.6
Marker Panel
Courtesy of Allied Signal General Aviation Avionics.

aircraft control and navigation. This occurs when it is impossible to see the ground on a continuous basis (due to rain, clouds, fog, etc.). Pilots, aircraft, and maintenance crews must abide by many rules in order to qualify for IFR flight. The instrument approach equipment must meet certain federal regulations. The technical guidelines concerning ILS facilities define these systems by Categories.

The ILS system provides three approach Categories that are determined by limitations based on visibility. The Categories indicate relative accuracy of the ILS guidance information:

- Cat I provides for accurate guidance to an altitude of 200 ft and a Runway Visual Range (RVR) of 2600 ft. This means that the pilot must have adequate external visual references at this distance and altitude in order to allow the landing to proceed.

- Cat II provides accurate guidance to an altitude of 100 ft and an RVR of 1200 ft.
- Cat III is subdivided into three parts:

  IIIA visual distance of 700 ft and 50 ft of altitude.

  IIIB visual distance of 150 ft and 35 ft of altitude.

  IIIC visual distance of 0 ft and 0 ft of altitude.

All categories legally allow the autopilot, not the pilot, to fly the aircraft down to the altitude limit. This is then referred to as "Decision Height." At this point the pilot either can or cannot see the runway to land. If the pilot cannot see the runway, he must do a go around, wait for weather to clear, or fly to an alternate destination. Note that the middle marker is located in such a position that if the aircraft is following the ideal glide path, then the middle marker is the place where the aircraft reaches Decision Height. It should also be noted that in Cat III operations the approach, landing, and rollout are all performed under automatic electronic control.

Runway visual range is a system that is located on the ground adjacent to the runway whereby a light source of known intensity is projected through a known distance (1200 ft), the electronic detector determines if the light is "visible" or not, and this information is transferred to the tower to aid the pilot in the decision-making process. RVR is reported to the nearest 100 ft.

For all categories of operation, the airport, the aircraft equipment, the flight crew, and the maintenance crews must be federally certified to the appropriate levels. Naturally Cat II must meet stricter standards than Cat I operation, and Cat III is the most stringent.

The Approach

When the decision to make an ILS approach is made, the pilot first must know how far he is from the airport. This kind of information was given to him from his en route navigation systems and should put his aircraft within 40 mi of the runway.

Localizer and glide slope are displayed to the pilot by means of deviation bars. These movable indicators are often referred to as **D Bars.** The marker information is displayed to the pilot by means of flashing lights and audio tones.

The **Course Deviation Indicator (CDI)** shown in Figure 1.7 will indicate when the aircraft is on the flight path. The localizer needle and the glide slope needle will be centered when the pilot has the aircraft on the ideal flight path. Navigation errors are represented by the displacement of the D Bar away from the centerline.

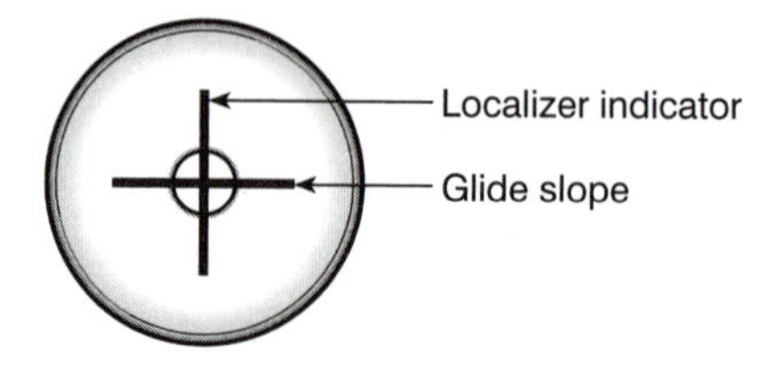

Figure 1.7
Course Deviation Indicator with Localizer and Glide Slope Deviation Bars

Use of the Localizer

When the aircraft is to the left of the ideal flight path the localizer needle will be to the right side of the instrument. When the needle is to the right side of the instrument it is indicating that the ideal flight path is to the right of the aircraft; the pilot should turn the aircraft slightly to the starboard side. *The pilot steers the aircraft toward the localizer needle* until the aircraft is again on course (see Figure 1.8).

The reverse is also true. When the aircraft is to the right of the ideal flight path the localizer needle will be to the left side of the instrument. When the needle is to the left side of the instrument it is indicating that the ideal flight path is to the left of the aircraft and the pilot should turn slightly to the port side. *The pilot steers the aircraft toward the localizer needle* until the aircraft is again on course (see Figure 1.9).

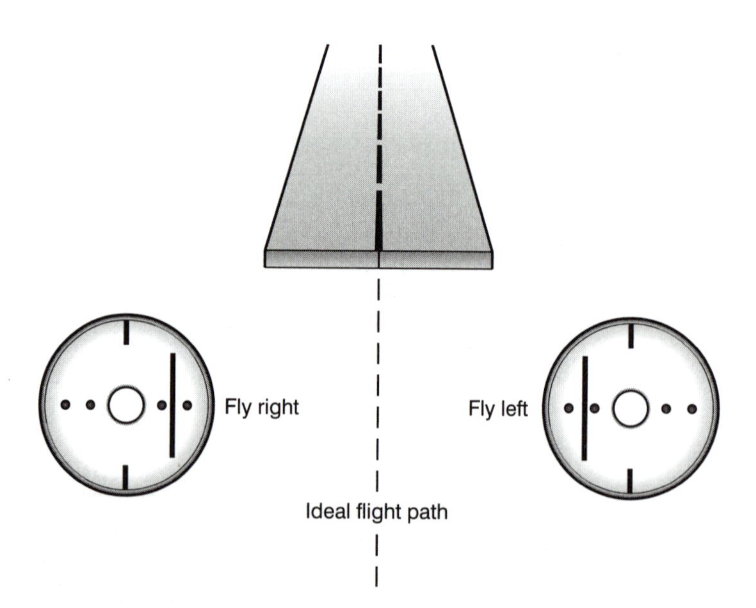

Figure 1.8
The CDI will be centered when the aircraft is on the ideal glide path.

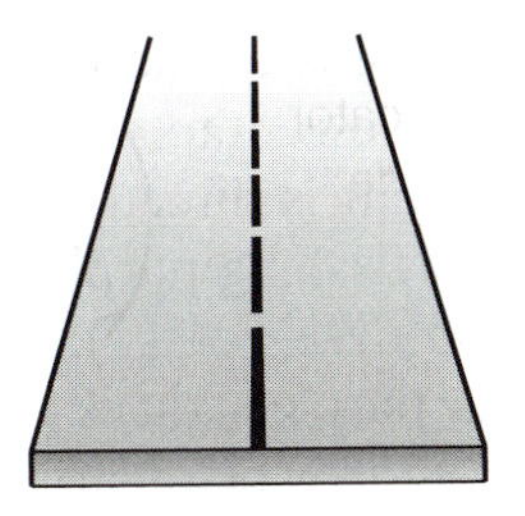

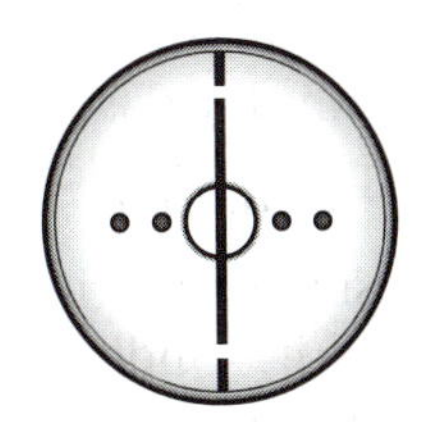

Figure 1.9
The CDI indicates when the aircraft is to the left or right of the ideal flight path.

Use of the Glide Slope

The glide slope needle works in a similar fashion to the localizer, but remember that the glide slope indicates the elevation parameter. When the needle is centered horizontally, the aircraft is on the ideal glide slope. When the needle is to the lower side of the instrument it indicates to the pilot that the ideal glide path is below the aircraft and he should then increase the rate of descent, thus putting the nose of the aircraft downward. *The pilot steers the aircraft toward the glide slope needle* until the aircraft is again on glide slope (see Figure 1.10).

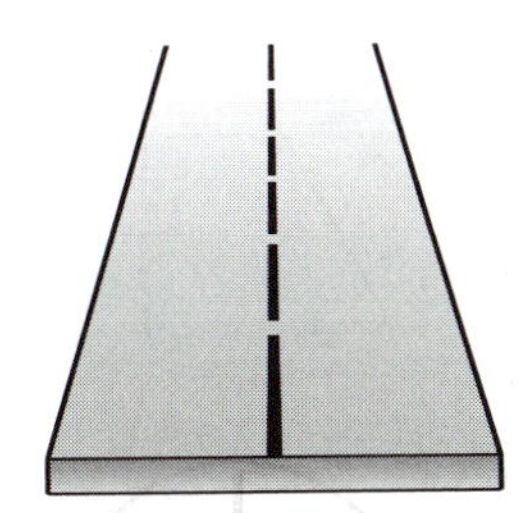

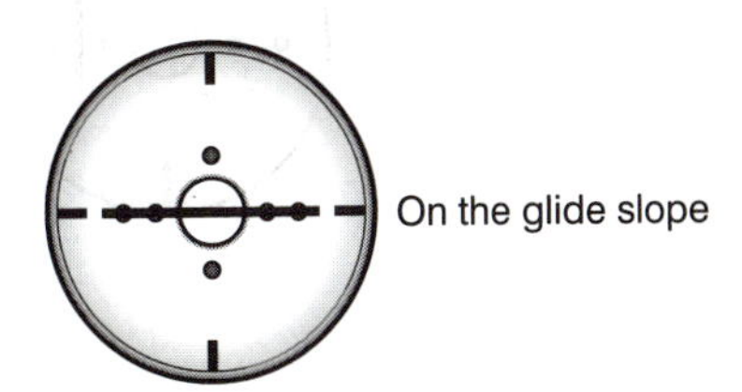

Figure 1.10
The CDI indicates when the aircraft is on the glide slope.

When the needle is to the top side of the instrument, the pilot should decrease his rate of descent to intercept the ideal glide path (see Figure 1.11).

When the aircraft is directly above the outer marker (about 6 mi out), the pilot will see a flashing purple (blue) light and will hear a 400 Hz tone that is coded dash dash (— —). The audio signal will sound through the audio system; the purple light is usually mounted on the right side of the instrument panel, or may be on an audio panel.

When the aircraft is directly above the middle marker (about 3500 ft out), the pilot will see an amber flashing light and will hear a 1300 Hz tone that is coded dot dash dot dash (•—•—). Remember that this is the decision height and the pilot must decide to land or abort.

ELECTRONICS

The localizer consists of three basic blocks—the receiver, the converter, and the indicator. Figure 1.12 shows their interconnection in a simplified diagram. The threefold task of the converter is

- to separate the 90 from the 150 Hz audio tones (see Figure 1.12)
- to compare amplitudes of both signals
- to measure the relative strength of the two audio signals.

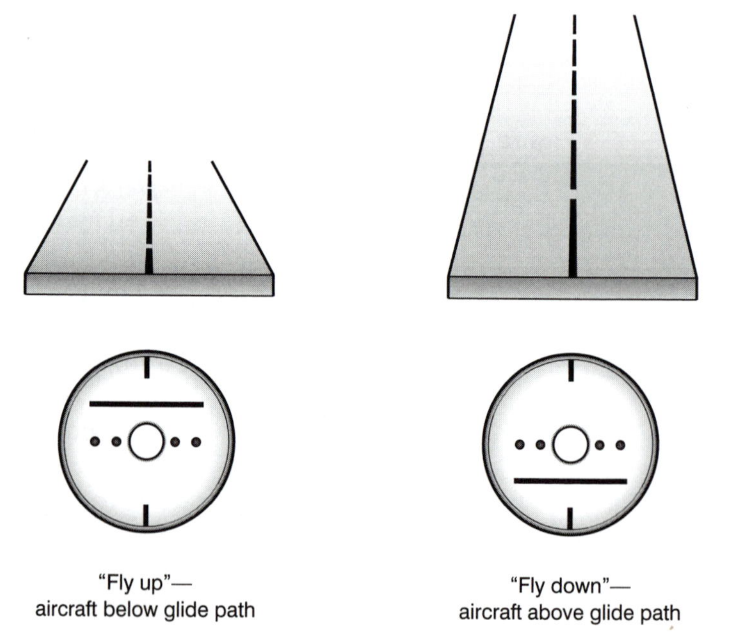

Figure 1.11
The CDI indicates when the aircraft is above and below the glide path.

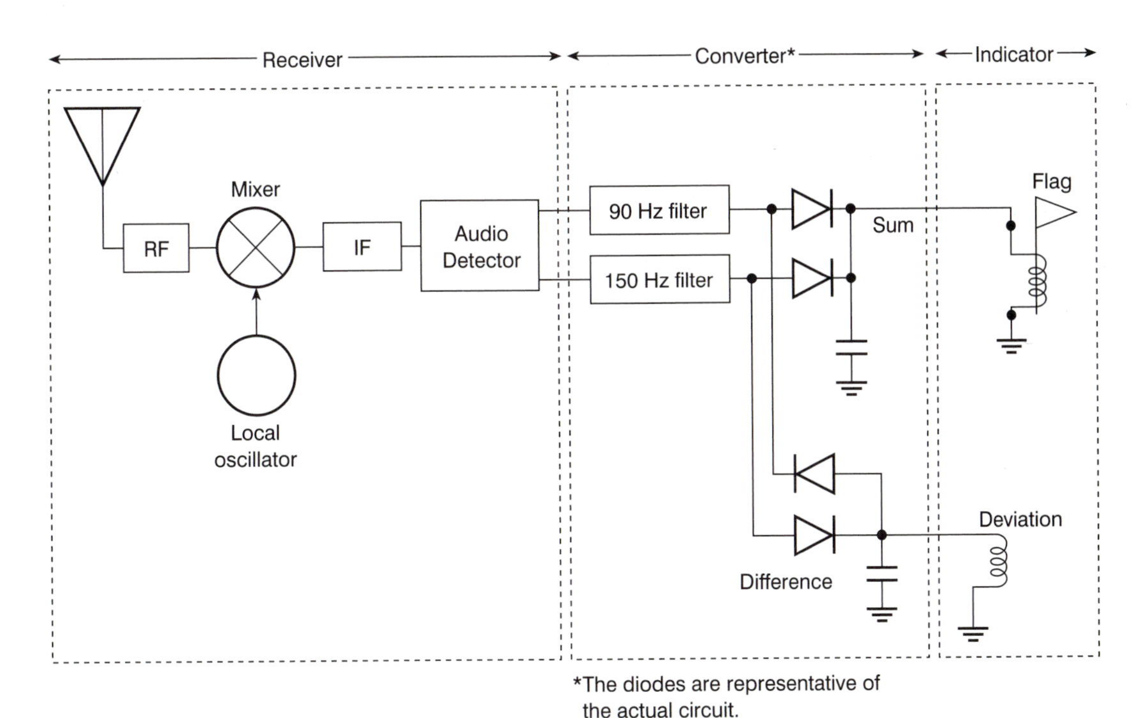

Figure 1.12
Simplified Block Diagram of the VHF Localizer

The first five blocks of Figure 1.12 describe any radio superheterodyne receiver. The detector output is then fed to the next set of blocks, which perform the task of the converter. The converter can be described as a set of filters and a voltage comparison diode network. The last two blocks of Figure 1.12 depict the indicator part of the system. The actual meter movements (Course Deviation Indicator) can be located in a CDI-style ILS indicator, a CDI-style localizer/VOR indicator, or an HSI.

The **ideal flight path** is defined where the two tones are of equal strength. The transmitter signal is amplitude modulated with 90 Hz and 150 Hz modulation, both at 30%. When the aircraft is located in the centerline of these two signals, and lined up directly with the runway, the signal strengths of the modulation are equal. The localizer needle is then centered. This occurs when the CDI current equals zero. A CDI has a standard amount of current (150 μA) required for full-scale deflection. This ensures that all CDI instruments operate at the same levels and provides for equipment compatibility and testing. These standards are established by the International Civil Aviation Organization (ICAO) and adhered to by manufacturers. A circuit called the Flag Driver measures the sum of the 90 Hz and 150

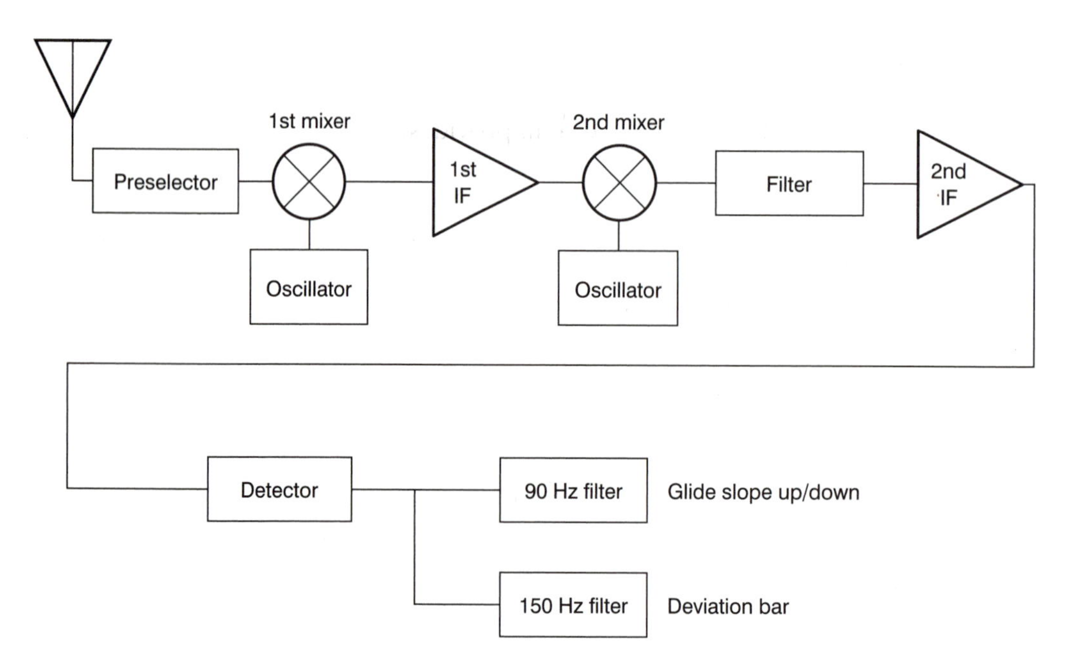

Figure 1.13
Simplified Block Diagram of the Glide Slope Receiver

Hz audio tones. If the navigation signals are not reliable or there are other faults in the receivers, then a warning flag is enabled to come into view. Both the localizer and the glide slope have Flag Driver circuitry, as represented by the diodes in Figure 1.12.

Figure 1.13 shows a block diagram similar to Figure 1.12, except that this one relates to the glide slope system instead of the localizer system.

Figure 1.14 shows a similar block diagram of the marker beacon receiver.

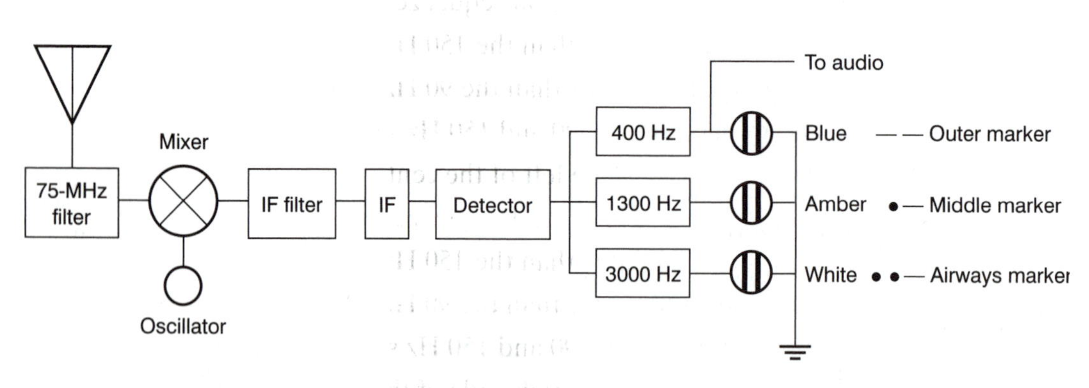

Figure 1.14
Block Diagram of the Marker Beacon Receiver

QUESTIONS

1. The instrument landing system provides:
   a. distance information
   b. azimuth information
   c. elevation information
   d. all of the above
2. The marker beacons provide:
   a. distance information
   b. azimuth information
   c. elevation information
   d. all of the above
3. The localizer provides:
   a. distance information
   b. azimuth information
   c. elevation information
   d. all of the above
4. The glide slope provides:
   a. distance information
   b. azimuth information
   c. elevation information
   d. all of the above
5. When the aircraft is centered on the centerline of the runway:
   a. the 90 Hz and 150 Hz signals equal zero
   b. 90 Hz signal is greater than the 150 Hz signal
   c. 150 Hz signal is greater than the 90 Hz signal
   d. the magnitudes of the 90 and 150 Hz signals are equal
6. When the aircraft is to the left of the centerline of the runway:
   a. the 90 Hz and 150 Hz signals equal zero
   b. 90 Hz signal is greater than the 150 Hz signal
   c. 150 Hz signal is greater than the 90 Hz signal
   d. the magnitudes of the 90 and 150 Hz signals are equal
7. When the aircraft is to the right side of the centerline on the runway:
   a. the 90 Hz and 150 Hz signals equal zero

b. 90 Hz signal is greater than the 150 Hz signal

c. 150 Hz signal is greater than the 90 Hz signal

d. the magnitudes of the 90 and 150 Hz signals are equal

8. What is the carrier frequency of the marker beacon?

a. 75 MHz

b. 108.1 MHz

c. 330 MHz

d. 400 Hz

9. Which indicating marker light will the pilot see in the cockpit when the aircraft flies over the outer marker?

a. white

b. blue

c. amber

d. red

10. Which indicating marker light will the pilot see in the cockpit when the aircraft flies over the middle marker?

a. white

b. blue

c. amber

d. red

11. What is the audio tone frequency of the middle marker?

a. 400 Hz

b. 1300 Hz

c. 75 MHz

d. 3000 Hz

12. What is the audio tone frequency of the outer marker?

a. 400 Hz

b. 1300 Hz

c. 75 MHz

d. 3000 Hz

13. A CDI has a standard amount of current required for full-scale deflection (established by the International Civil Aviation Organization). The amount of current required for full-scale deflection is:

a. 50 μA

b. 78 μA

c. 90 μA

d. 150 μA

14. When the NAV flag is in view on a CDI:
    a. one of the modulating tones of the localizer is missing
    b. the AGC (Automatic Gain Control) of the receiver is responding to a weak signal
    c. the navigation information is normal
    d. the navigation information is not reliable
15. A pilot who intends to land on the back course approach must remember:
    a. the CDI has reverse needle sensing
    b. on the approach map the shaded area is on the left side
    c. there is no glide slope
    d. all of the above
    e. the wind will now be from behind the plane
    f. none of the above
16. Match the carrier frequency with the correct navigation system.

    112.3 MHz   Localizer ____________________

    334.7 MHz   Marker beacon ____________________

    75 MHz   Glide slope ____________________
17. Sketch the radiation pattern of a standard ILS (three parts).

CHAPTER 2

Microwave Landing System

MLS

Introduction

This chapter deals with the basic principles of the Microwave Landing System (MLS). The MLS under development was scheduled to replace the current ILS within a decade.

Objectives

Explain the basic principle of operation of the MLS

Describe the subsystems required for basic MLS operation

Describe the azimuth and vertical coverage angles

Sketch the transmitter antenna locations on an airport

Identify the operational frequencies and channels

List the number of channels in MLS

State the channel increments through the frequency band

Explain what the pilot sees and what she can control

Identify the basic system block diagram

State the time required to transmit an MLS data transmission

Identify the component parts of the digital data transmission

Basic Concept

The Microwave Landing System (MLS) is an electronic aid to navigation that will helps the pilot to position the aircraft for landing in conditions of reduced visibility.

MLS THEORY

Microwave Landing System or MLS was designed to replace the existing but more rigid Instrument Landing System. MLS is an electronic aid to navigation that assists the pilot in positioning the aircraft for landing in conditions of reduced visibility. Some functional MLS sites are already in use during the development phase.

The use of a scanning microwave beam will make instrument landings more accurate and flexible. MLS allows curved approach paths. The system can operate on 200 channels in the C band, and is called time-referenced scanning beam or TRSB. The curved approach paths enable the aircraft to shorten its direct approach to avoid populated areas. MLS will also reduce delays and noise and increase safety standards by providing airports with Category upgrades. Because it operates in the microwave frequency band, MLS is not subject to interference and multipath (or siting) errors.

The current generation of ILS, in contrast, is subject to five major problems:

- sitting errors,
- multipath errors,
- limitation in the number of available channels (40),
- radio frequency (RF) interference (man-made or environmental noise), and
- a requirement of long, narrow, straight-in approach paths.

MLS will eliminate all five problems.

In addition, installation and maintenance costs for MLS will be much less than for an ILS installation, although there is a requirement for a DME/P interrogator that is just now being developed. Due to the increased number of channels available with MLS, the installation of MLS facilities will likely mushroom in heavy air traffic areas. Often, larger airports have more than one instrument runway, so there may be two or more localizers and glide slopes for every large airport, and one for every feeder airport. This increases the chance of interference from one ILS to the next. MLS will eliminate this problem.

Another advantage of MLS is that it is adaptable to every airport size and geographic location. From the smallest airport, which will use the basic system, to the largest, which will install all of the MLS features, aircraft will have the same flexibility with this system. Airlines can install in their aircraft only those features of the MLS that their particular operation requires. Smaller aircraft with fewer MLS features will be able to utilize large airports, and larger aircraft, which may be fully equipped, will be able to make instrument landings at smaller airports using only those MLS facilities available at that station. The Federal Aviation Administration was scheduled to make MLS implementation decisions by 1995.

PILOT PERSPECTIVE

What does the pilot see when using MLS? She sees standard course deviation indicator (CDI) and/or horizontal situation indicator (HSI) information (azimuth, elevation, and distance).

ELECTRONICS OPERATION

Each MLS installation uses only one of 200 available frequencies, spaced every 300 kHz, for the azimuth and elevation portion of the system, and operating between 5031 and 5090 MHz in the C band.

Frequency	Channel
5031 MHz	500
5031.3	501
5031.6	502
5031.9	503
5032.2	504
etc.	etc.
5090.7	699

The **Distance Measuring Equipment (DME)** portion of the MLS uses one of the regular DME channels in the L band, from 978 MHz to 1215 MHz. The basic MLS contains these functions:

- approach azimuth
- approach elevation
- DME
- available space for enhancement

Expanded MLS may contain all of the above, and in addition, can contain

- back azimuth
- flare elevation
- back DME

Figure 2.1 shows MLS antenna locations.

Approach Azimuth and Elevation Scanning Beam Coverage

Figure 2.2 shows an approach lateral azimuth of 0° to ±40°.

Figure 2.3 illustrates that TO and FRO scanning beams move counter-clockwise and clockwise.

Figure 2.4 gives a side view of the approach elevation scanning beam.

Both azimuth and elevation limits can be expanded. The azimuth can be expanded to ±60° and the elevation can be expanded to 30°.

By means of electronically scanning the antennas, the azimuth and elevation information is imparted to the receiver in the aircraft. Both azimuth and elevation have their own TO and FRO scanning beams. For simplicity only the top view of the azimuth antenna is shown in Figure 2.3. Directional information is transmitted by microwave.

Figure 2.1
MLS Antenna Locations—Satellite View

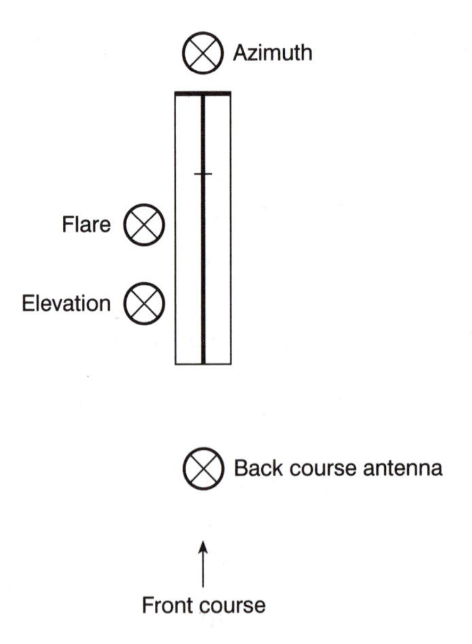

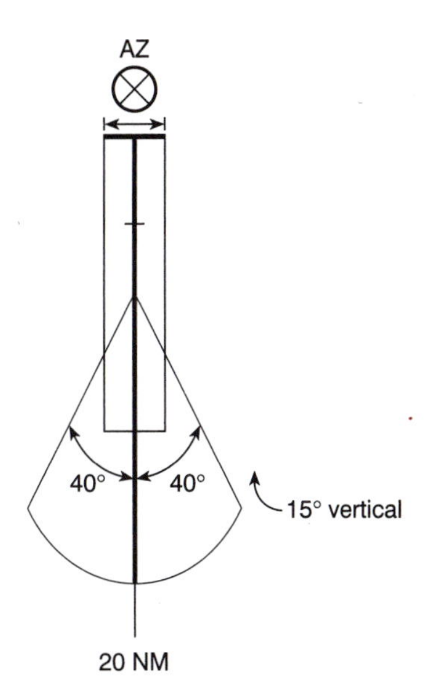

Figure 2.2
Top View of the MLS Approach Azimuth Scanning Beam

The aircraft receiver measures the time between the TO and FRO scans, and is then able to determine the aircraft's position. Aircraft often have two antennas, one located forward and one aft on the belly centerline.

Data Transmission Formats

Transmissions are encoded by differential phase shift keying (DPSK) at the start of each transmission. Rapidly switching the phase of the microwave car-

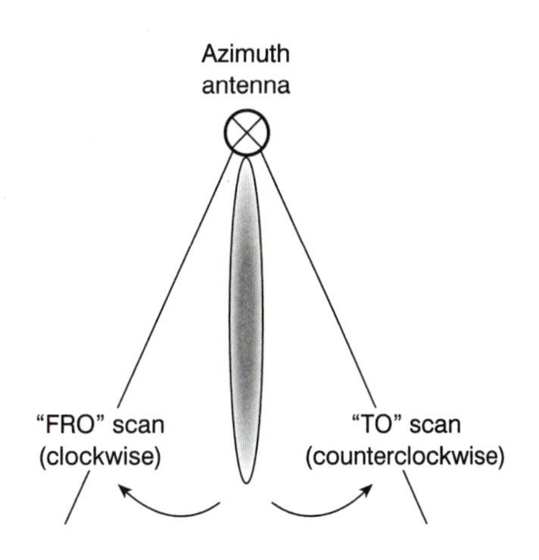

Figure 2.3
The "TO" scanning beam operates counterclockwise and the "FRO" scanning beam clockwise.

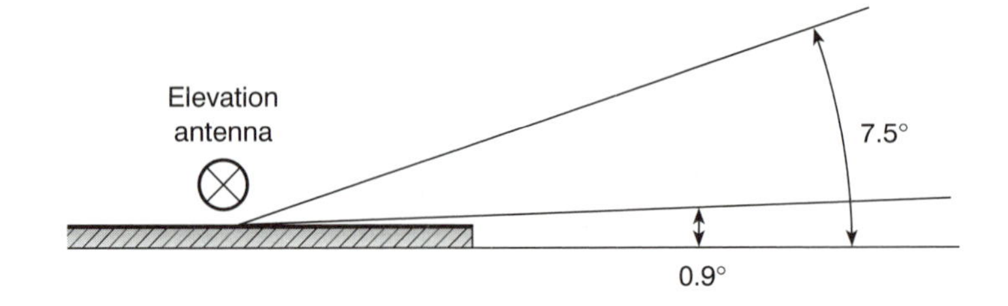

Figure 2.4
Side View of the MLS Approach Elevation Scanning Beam

rier to send digital data creates DPSK. The logic one is obtained by rapidly phase shifting 180°, and the logic zero is obtained by no phase shift. MLS is a time-share system, each function transmitting a data block in turn. The data block looks like Figure 2.5.

Preamble Contents

Each data block starts with a **preamble.** The preamble uses differential phase shift keying, and organizes the aircraft hardware for communication to occur. Preamble and sector signals, as shown in Figure 2.6, assure that the aircraft is receiving the signal and is in the correct mode to use the received data. DPSK measures phase changes; therefore, the airborne system must phase lock to the transmitted signal.

Components of the preamble signal are

- **Radio frequency (RF)** carrier acquisition period: An unmodulated RF carrier that will allow the airborne receiver to phase lock.
- Receiver reference time code: A timing reference point against which data bits are measured.
- Function identification code: Identifies the specific parameter or function that is to be received, such as azimuth, elevation, or flare.

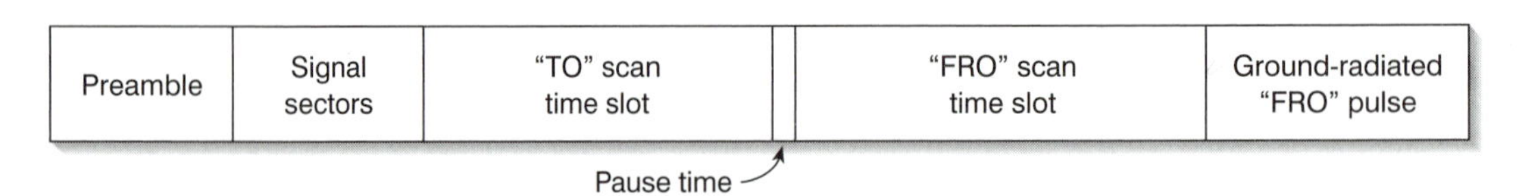

Figure 2.5
MLS Data Transmission Sequence

Figure 2.6
Preamble Contents

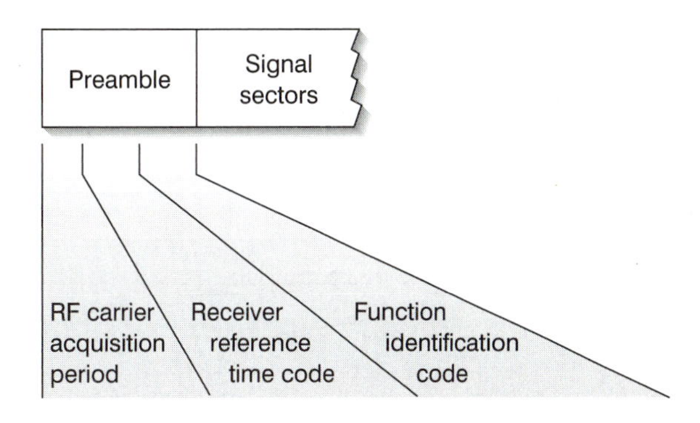

Sector Signal Contents

Each preamble is followed by a sector signal, as shown in Figure 2.7, components of which are

- Ground equipment identification: A digital command to turn on or off an audio oscillator that will produce a Morse code airport identifier.
- Airborne antenna selection: A long unmodulated carrier to allow testing for the strongest incoming signal from either the front or the rear antenna.
- OCI (Out of Coverage Indication): When an aircraft is out of the scanning zone of the MLS, it will receive the left, right, or rear OCI signal. The indication to the pilot depends on which signal is received.
- "TO" test pulse: A pulse that tells the receiver that the "TO" signal is about to happen.

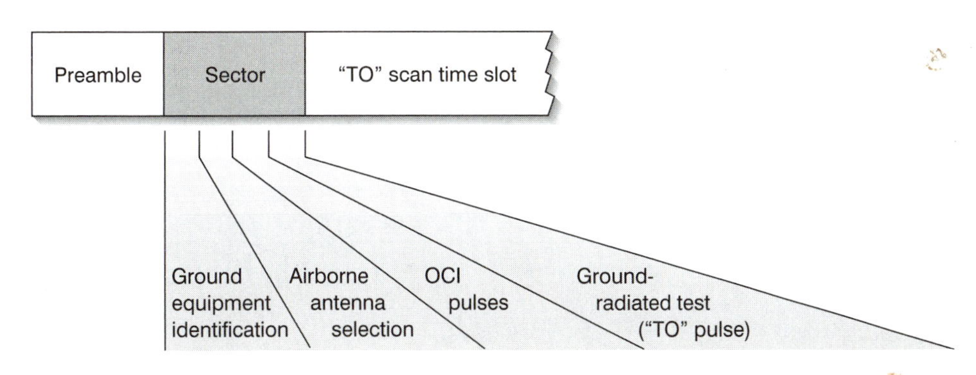

Figure 2.7
Sector Signal Contents

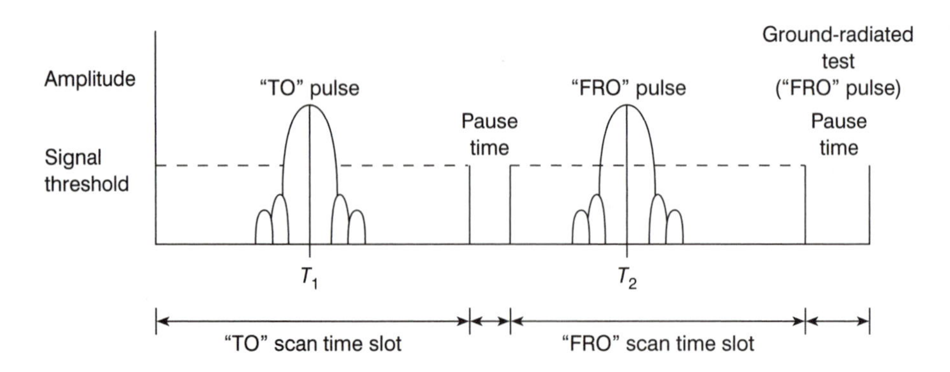

Figure 2.8
Organization of the TO and FRO Scan Time Slots

TO and FRO Scan Time Slots

Figure 2.8 explains how the **TO and FRO scan** time slots are organized and Figure 2.9 shows the sequence in which antenna digital data are transmitted. Each sequence contains four parts:

- elevation
- flare

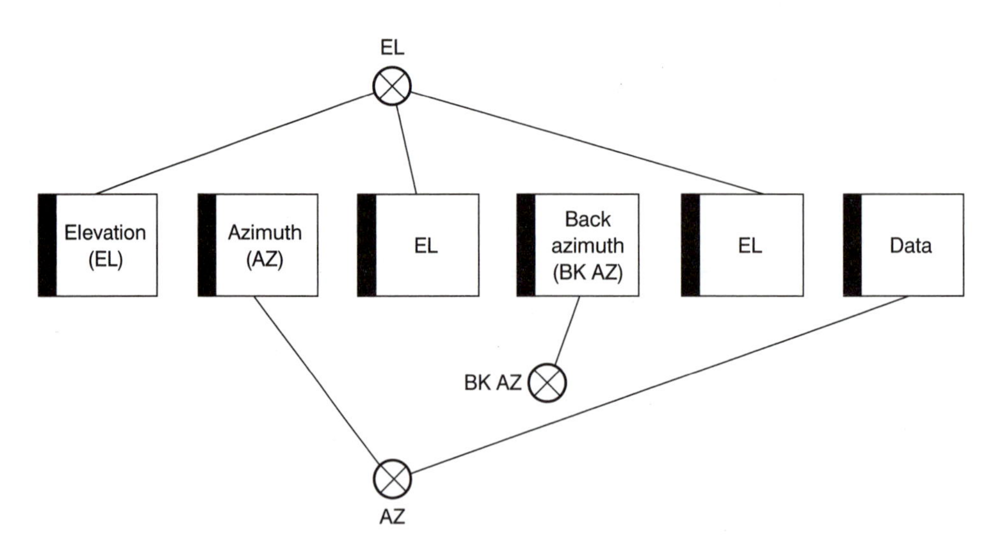

Figure 2.9
Antenna Digital Data Transmission Sequence

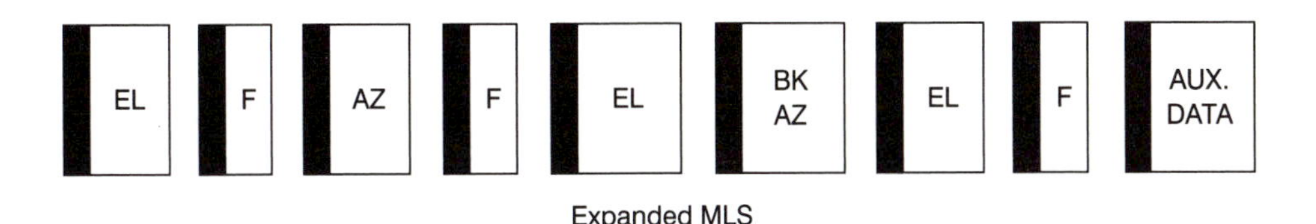

Figure 2.10
Expanded MLS Transmission Sequence

- azimuth
- back azimuth

The sequences are transmitted twice sequentially for a total time not to exceed 134 ms.

Each data element in the sequence is allocated a certain position in the transmission sequence, as seen in Figure 2.10. A full cycle of transmission in MLS consists of four groups, with each group containing two sequences, for a total of 615 ms.

QUESTIONS

1. What is the azimuth angle of the approach?
2. What is the elevation angle for the approach?
3. How many channels are available with the MLS?
4. What is the frequency range for MLS?
5. Name two major advantages of MLS over ILS.
6. What are the four major functions required of basic MLS installations?
7. Sketch the transmission sequence of a digital data transmission, and identify all the portions of the transmission.
8. Identify the azimuth and elevation antenna locations on an airport sketch.
9. Where are the optimal locations for MLS antennas on an aircraft?

CHAPTER 3

Radio Magnetic Indicator

RMI

Introduction

This chapter deals with the basic principles of the Radio Magnetic Indicator (RMI).

Objectives

Explain the general function of the RMI

Explain the function of the compass card

Explain the function of the number one needle

Explain the function of the number two needle

Determine the compass heading of the aircraft

Determine the direction to selected stations

Sketch a top map view of the aircraft and the station position

Basic Concept

The Radio Magnetic Indicator is primarily used as a nonprecision aid to enroute navigation, although it can be useful on an ILS approach.

THE MULTIPURPOSE RMI

The **Radio Magnetic Indicator** is used primarily as a nonprecision aid to enroute navigation, although it can be useful on an ILS approach. This instrument is a multipurpose unit that has:

- two pointers (needles)
- compass card
- compass warning flag

The RMI receives information from the Compass System, the **VOR,** and/or the **ADF** systems. Both portions can operate independently of each other. The RMI displays compass information and either automatic VOR or ADF bearing. The selection between these two options is accomplished by a switch on the face of the instrument for each of the two indicator needles.

One indicator needle is a single bar needle and the other is a double bar needle. The single needle is the one with the thin single metal pointer. The single bar needle presents information from the number 1 system, either VOR or ADF. The double bar needle is the one with the two parallel pointers. The double bar needle presents information from the number 2 system, again either VOR or ADF. Figure 3.1 shows the instrument.

The compass information is received from repeated heading information received from the Compass System. This information is often referred to as "slaved Compass" information.

Figure 3.1
Radio Magnetic Indicator
Courtesy of Allied Signal General Aviation Avionics.

QUESTIONS

1. Compass information is displayed by:
   a. the number 1 needle
   b. the number 2 needle
   c. the compass card
   d. the flux valve
2. VOR bearing information is displayed by:
   a. the number 1 needle
   b. the number 2 needle
   c. the compass card
   d. either number 1 needle or number 2 needle
3. ADF bearing information is displayed by:
   a. the number 1 needle
   b. the number 2 needle
   c. the compass card
   d. either number 1 needle or number 2 needle
4. The RMI can display:
   a. VOR information
   b. ADF information
   c. compass information
   d. all of the above
5. From the RMI shown in Figure 3.2, number 1 needle indicates:
   a. 11°
   b. 36°
   c. 98°
   d. 134°
   e. 210°
   f. 269°
   g. 360°
6. From the RMI shown in Figure 3.2, number 2 needle indicates:
   a. 11°
   b. 36°
   c. 98°

Figure 3.2
RMI

d. 134°
e. 210°
f. 269°
g. 360°

7. What is the heading of the aircraft?
   a. 11°
   b. 36°
   c. 98°
   d. 134°
   e. 210°
   f. 269°
   g. 360°

CHAPTER 4

Horizontal Situation Indicator

HSI

Introduction

This chapter deals with the operational principles of the Horizontal Situation Indicator (HSI).

Objectives

Explain the general function of the HSI

Identify the three major sections of the indicator

Compare the HSI with the RMI

Sketch a top map view of the aircraft and relative station position

Basic Concept

The Horizontal Situation Indicator is a multipurpose navigation instrument.

THE HSI AS A MULTIPURPOSE NAVIGATION INSTRUMENT

The **Horizontal Situation Indicator (HSI)** (seen in Figure 4.1) is a multipurpose navigation instrument. This instrument receives information from a variety of sources and provides the flight crew with information from:

- localizer
- glide slope
- VOR
- slaved Compass

In addition, the indicator provides a variety of warning and supplemental information such as:

- TO/FROM arrows
- aircraft heading
- warning flags
- heading bug

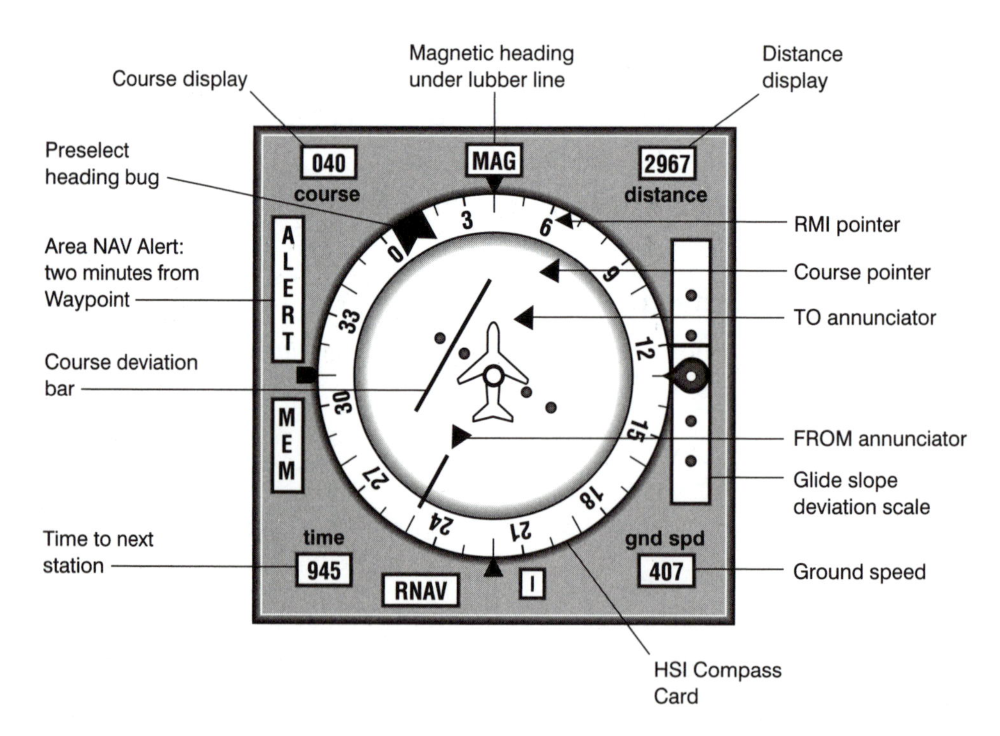

Figure 4.1
Horizontal Situation Indicator

The HSI receives information from the Compass System, the ILS, and the VOR. Various portions of the instrument can operate independently of each other. Naturally the purpose of the instrument is to make navigation easier for the flight crew, and therefore all information is preferred. The HSI indications are not useful without compass information. The compass information is received from repeated heading information received from the Compass System. This information is often referred to as "slaved Compass" information.

The VOR information is displayed and adjusted very much like the Course Deviation Indicator, with the exception of the course select knob, which replaces the **omni bearing select (OBS)** knob. The heading bug is used to remind the pilot about the next heading to which she is to fly. In more complicated autopilot systems the heading bug actually provides an input to the autopilot.

The advantage of an HSI is that it displays the actual spatial relationship between the aircraft and the desired course as if viewed from a satellite. The deviation bar swings left and right, but is also free to rotate with the compass card. If the intercept to the desired course is 30°, then the angle between the symbolic aircraft and the course deviation bar is also 30° (see Figure 4.1). Very commonly, an HSI will also contain an RMI pointer associated with the compass card indication. This operates in the same way as a conventional RMI, and may be used to indicate either automatic VOR or ADF bearing information.

QUESTIONS

1. Compass information is displayed by:
   a. course pointer
   b. the number 2 needle
   c. the course selection knob
   d. compass card
2. Pilot selected VOR radial information is displayed by:
   a. the course pointer
   b. the number 2 needle
   c. the compass card
   d. glide slope flag
3. ADF bearing information is displayed by:
   a. the lubber line

b. the course select needle
c. the compass card
d. the little airplane
e. the RMI pointer

4. The HSI can display:
   a. VOR information
   b. localizer information
   c. glide slope information
   d. all of the above
5. From the HSI shown in Figure 4.2, course pointer indicates:
   a. 20°
   b. 40°
   c. 70°
   d. 130°
   e. 220°
   f. 250°
   g. 310°
6. According to the preselect heading bug in Figure 4.2, the next course to steer is:
   a. 10°
   b. 40°
   c. 70°
   d. 130°
   e. 220°
   f. 250°
   g. 310°
7. What is the heading of the aircraft according to Figure 4.2?
   a. 20°
   b. 40°
   c. 70°
   d. 130°
   e. 220°
   f. 250°
   g. 310°

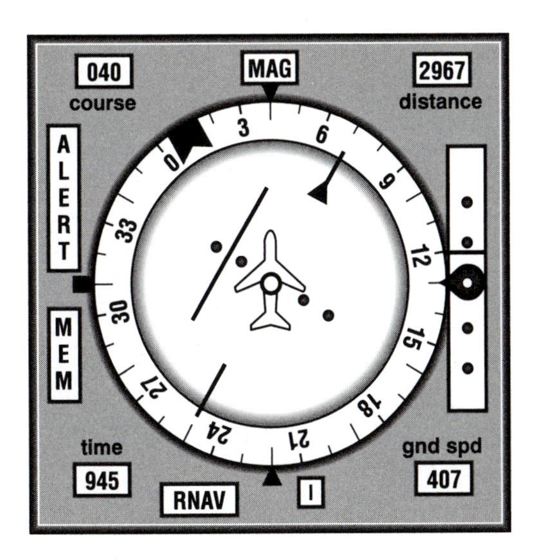

Figure 4.2
HSI

8. Based on Figure 4.2, if the pilot has selected a VOR station, what radial has she selected?
   a. 20°
   b. 40°
   c. 70°
   d. 130°
   e. 220°
   f. 250°
   g. 310°
9. Based on Figure 4.2, if the pilot has selected a VOR station, is the selected course left or right of the current flight path?
   a. left
   b. right
   c. pilot is on the correct flight path
10. Based on Figure 4.2, if the pilot was on an ILS approach, what direction would she fly?
    a. left
    b. right
    c. up
    d. down
    e. left and up
    f. right and down

CHAPTER 5

Very High Frequency OMNI Range

VOR

Introduction

This chapter deals with the basic principles of Very High Frequency (VHF) navigation systems.

Objectives

Identify the basic principles of operation

Describe the radiation pattern of the VOR signal

Describe the modulation components of the VOR signal

Identify the frequency range of operation

Explain what is meant by reference signal

Explain what is meant by variable signal

Identify the block diagram of the VHF system

Explain the function of the omni bearing selector

Explain the function of the TO/FROM indicator

Describe the antenna for the VOR

Basic Concept

VOR is an acronym for **V**ery High Frequency **O**mni **R**ange. VOR is a navigation aid used to indicate the direction to fly, in order to reach a given ground station. VOR is used primarily for enroute navigation.

BASIC CONCEPT OF VOR

The acronym VOR stands for ***V*ery High Frequency *O*mni *R*ange,** which is a navigation aid used to indicate the direction to fly in order to reach a given ground station. VOR is used for enroute navigation.

Low frequency and medium frequency signals suffer from atmospheric static, precipitation static, and night effects. The advantages of VHF radio navigation is its relative immunity to propagation and atmospheric effects. VHF is essentially line of sight with no ionospheric propagation. Maximum range of VHF is a function of altitude, typically up to 150 NM.

The purpose of the VOR is to

- provide a means to determine an aircraft's position with reference to a VOR ground station
- provide a certain path to follow, toward or away from, a VOR station.

This is accomplished by indicating when the aircraft is on a selected VOR station **radial,** or by determining which radial the aircraft is on. The VOR indicates the aircraft's relative position with respect to the radial selected. VOR stations are charted on aeronautical charts and airport guides. To determine which radial the aircraft is on, compare the phase difference between signals. These signals are generated by the VOR ground station.

Figure 5.1 shows three radials (210, 80, 30). The more easterly one is 080°, indicating that the aircraft is (at a radial of) 080° magnetic with respect to the station. The aircraft could actually be flying in any direction. All directions are referenced to magnetic north.

If the aircraft is on the 210 radial, we say that the station's **bearing** is 030, which is the magnetic heading the pilot would fly if he wished to fly

Figure 5.1
Radials and Bearings

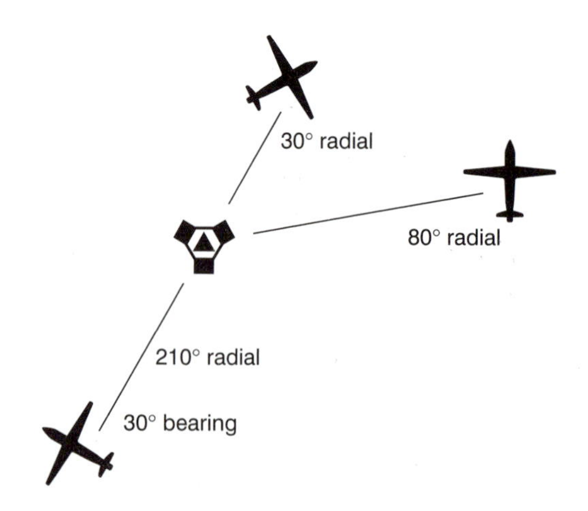

toward the station. The actual heading of the aircraft is of no concern to the VOR system.

When the aircraft is flying directly over top or alongside the VOR station, the TO/FROM flag will change position and the course deviation bar will flutter or act "nervous." These are the indicators that in fact the aircraft is near and passing the station.

Note: The radial is always away from the station, and the bearing is always toward the station.

VOR Analogy

1. Consider a directional beam of white light that is set up so that it rotates through 360° at a constant speed.
2. Whenever the white directional light points north, a red light flashes instantaneously in all directions (omnidirectional).
3. By knowing the time of rotation of the directional white light, and the space of time between the white and red flashes, observers can determine their direction from the station.

PILOT PERSPECTIVE

The pilot has control of the VOR system

- by channel selection on the NAV number 1 through the frequency range of 108.00 to 117.95 MHz.
- by adjusting the Omni Bearing Selector knob (OBS), which allows the pilot to rotate the course select card (or bearing select card) and thus select a given radial.

The pilot receives information from the VOR system through

- the Course Deviation Indicator (CDI), or zero centering meter, which indicates the aircraft's relative position with respect to the selected radial. (The selected radial is established by the OBS knob.)
- the white TO/FROM flag, which indicates either toward the selected radial or away from the selected radial.
- the NAV flag (usually red or barber pole), which indicates when the signal is below the acceptable level for the equipment to function correctly.
- audio—1000 Hz Morse code or a voice signal for station identification.

It should be noted here that as the pilot flys directly over a VOR, the signal becomes very confused and the CDI will slew quickly (scallop), the flag may pop into view, and the TO/FROM will jitter. This is known as the **Cone of Silence** or **Zone of Confusion.**

COURSE DEVIATION INDICATOR AND HORIZONTAL SITUATION INDICATOR

The CDI (Course Deviation Indicator, illustrated in Figure 5.2), or in a more sophisticated system the HSI (Horizontal Situation Indicator, shown in Figure 5.3), will provide the pilot with left/right information on the course deviation bar and TO/FROM indication. The indicator also has control over the course selection by the course selector pointer or Omni Bearing Selector (OBS). The course selector knob may be mounted in the lower left corner of the HSI unit or mounted remotely.

COURSE DEVIATION INDICATIONS

Figure 5.4 shows the course deviation indications for various positions. Note that these indications are independent of aircraft heading. For example, when the aircraft is south of the station (position E) and OBS is set to zero,

Figure 5.2
Course Deviation Indicator

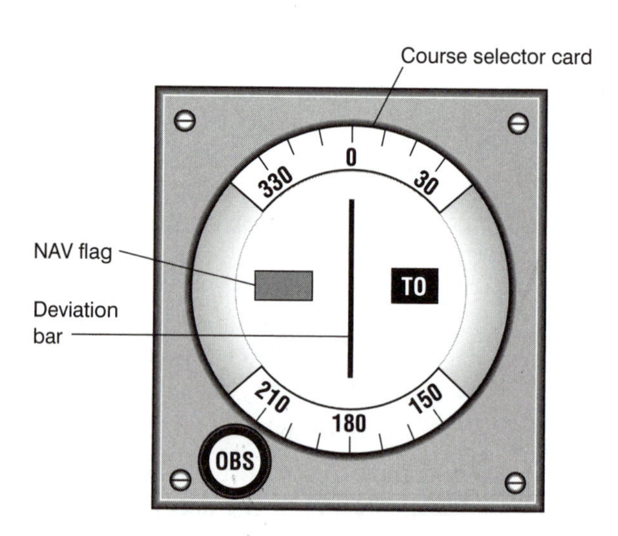

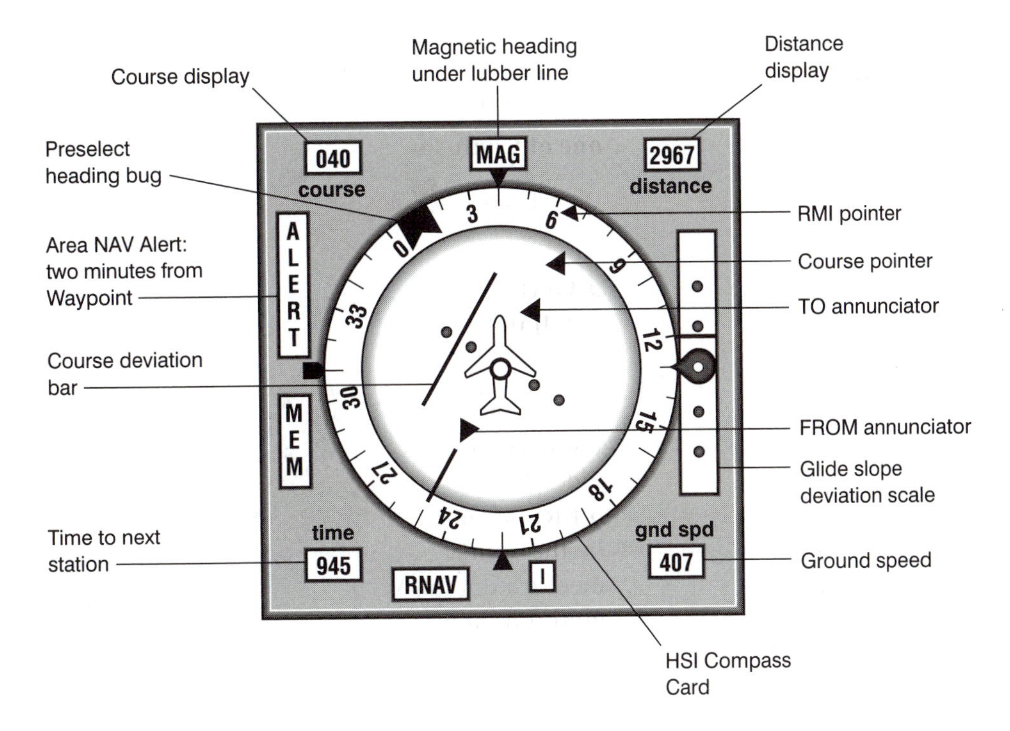

Figure 5.3
Horizontal Situation Indicator

then the needle is centered and a "TO" indication shows, but when the aircraft is north of the station, in position A, then the CDI shows a "FROM" indication and a centered needle.

VOR PRINCIPLES OF OPERATION

Carrier Frequency

At this point, let's look at the carrier frequency arrangement before discussing the actual VOR operation. The carrier frequency of the VOR is

- from 108.00 to 112.00 MHz on all of the even 100 kHz at 50 kHz increments
- from 112.00 to 117.95 MHz at 50 kHz increments on both odd and even 100s

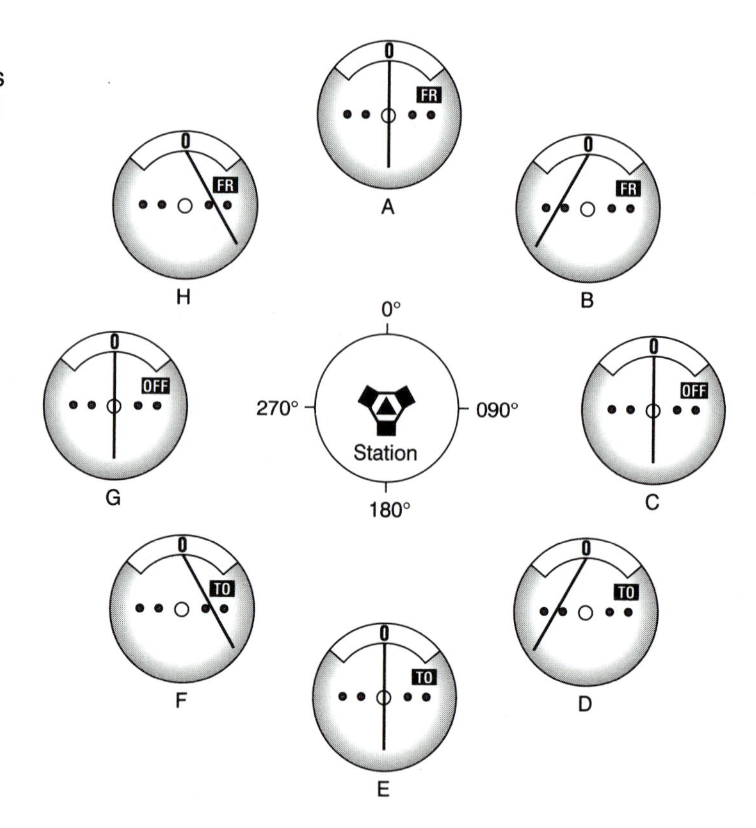

Figure 5.4
Course Deviation Indications for Various Positions Around the Compass

VOR uses the same frequency band as the localizer, and in fact, uses the same antenna and parts of the same receiver (108.00 to 117.95 MHz). Common systems are the antenna, receiver, and detector.

The frequency spectrum looks like this:

108.00	Reserved for test purposes
108.05	VOR
108.10	Localizer
108.15	Localizer
108.20	VOR
108.25	VOR
108.30	Localizer
108.35	Localizer
etc.	

See Appendix A, Radio Frequency Spectrum.

Types of Signals Generated

Now we can discuss VOR operation. The VOR ground station generates two signals that are riding on the carrier frequency. Let's assume that the pilot has tuned the navigation receiver to 108.05 MHz. This carrier will have three signals impressed on it, as follows:

1. The first signal will be an amplitude modulated signal called the **Variable signal.** The VOR ground station rotates the Variable signal at 1800 rpm, which will appear as a 30 Hz **amplitude modulated signal** to a receiver at one specific radial.
2. The second signal will be a 9960 Hz amplitude modulated signal. The 9960 AM waveform will then have a frequency modulated 30 Hz **Reference signal.** The Reference signal is modulated this way in order to keep the Reference and Variable 30 Hz signals separated until they are compared in the converter circuit.
3. The third is amplitude modulated Morse code for station identification.

The Reference and the Variable signals are in phase, at 0° magnetic. The Reference signal has the same phase in all directions from the ground station at any given instant in time, like the red light in the analogy. The Variable signal varies in phase as the direction from the VOR station increases clockwise, like the white light in the analogy. For example, if the aircraft is on the 030° radial, the Variable signal will lag the Reference signal by exactly 30°. By comparing the phase difference between the Variable and the Reference signals, the aircraft bearing relative to the VOR station can be determined.

Modulation Levels

The modulation levels of a VOR ground station transmission are:

- 30 Hz at 30%
- 9960 Hz at 30%
- 1020 Hz at 30%

The major disadvantage of this multimodulation and multiple signal is that the relative strength of the variable signal changes as the antenna changes its position, creating multipath errors.

The VOR is identified on aeronautical charts with a small box that contains information about the VOR. The identifier of the VOR or **VORTAC** is the station name, the frequency in MHz, and the three-letter Morse code identifier for the station, as shown in these examples.

VOR
STILLWATER
108.4 SWO

VORTAC
KANSAS CITY
112.6 CH 73 MKC

VOR Converter

There are several ways to retrieve the VOR information. The signal processing for most equipment is similar, as shown in Figure 5.5.

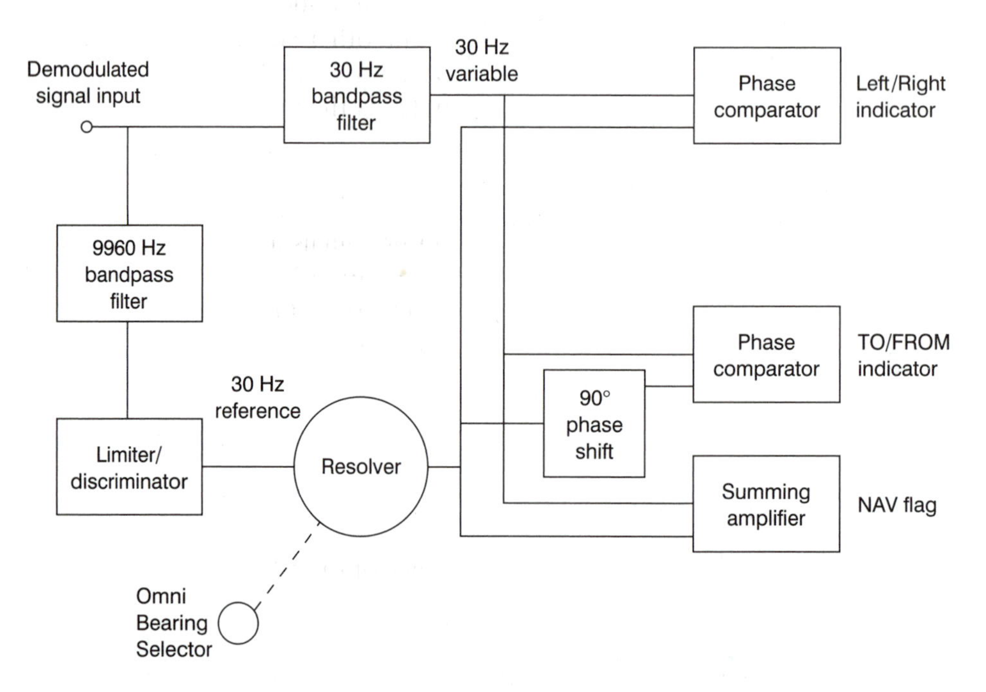

Figure 5.5
VOR Converter Block Diagram

1. First separate the two 30 Hz signals.
2. Phase shift one of these signals through the resolver (OBS)—usually the reference signal.
3. Compare the two phases of the two signals and then use the resultant signal to drive the left/right bar.
4. Apply a 90° phase-shifted signal to the resolver output, and compare this signal to the variable signal; then use this to drive the TO/FROM flags.

Ground Station Signal Generation

There are two ways for the ground station to generate the complex resultant cardioid signal.

1. The first method is called the **C VOR,** C standing for conventional. The conventional method uses Alford loops to electronically spin the variable signal.
2. The second method is called **D VOR,** D standing for Doppler. The Doppler method uses a 30 Hz amplitude modulated signal generated in an omnidirectional pattern. The other signal is produced by generating two frequencies; one, 9960 Hz above the carrier; and the other, 9960 Hz below the carrier. This complex is then sent to an array of 52 antennas mounted on the ground in a 44 ft circle so as to make the signal appear to rotate at 1800 rpm.

Both the localizer and the VOR signals are processed by the NAV converter. The VOR compares phase relationships of the 30 Hz Reference and Variable signals. The antenna used by the VOR and the localizer is generally a 0.25 wave horizontally polarized type.

QUESTIONS

1. What is the reference frequency of the VOR?
   a. 30 Hz
   b. 9960 Hz
   c. 1020 Hz
   d. 108.15 MHz

2. What is the variable frequency of the VOR?
   a. 30 Hz.
   b. 9960 Hz
   c. 1020 Hz
   d. 108.15 MHz
3. What is the subcarrier frequency of the VOR?
   a. 30 Hz
   b. 9960 Hz
   c. 1020 Hz
   d. 108.15 MHz
4. The reference frequency of the VOR is:
   a. amplitude modulated
   b. frequency modulated
   c. phase modulated
   d. nonexistent
5. The VOR uses the same radio receiver as the localizer. (True or False?)
6. The OBS (Omni Bearing Selector) rotates a bearing selector card. (True or False?)
7. When flying on a heading toward a VOR station with a "TO" indication (compass heading agrees with OBS), and the CDI needle is to the left side of the indicator:
   a. fly aircraft to the left
   b. fly aircraft to the right
   c. maintain speed and altitude
   d. maintain heading

CHAPTER 6

Automatic Direction Finder

ADF

Introduction

This chapter deals with the basic principles of the Automatic Direction Finder.

Objectives

Identify the basic concept of ADF

Identify the frequency range of operation

Explain how the unit functions

Explain how radio frequency (RF) is propagated

Describe the ADF loop antenna reception pattern

Describe the ADF sense antenna reception pattern

Describe the resultant cardioid reception pattern

State optimum sense and loop antenna location on aircraft

Analyze the block diagram of typical ADF

Describe the bidirectional antenna pattern

Describe the omnidirectional antenna pattern

State audio outputs from ADF

Basic Concept

The Automatic Direction Finder is a navigation aid that provides relative bearing information.

ADF CONCEPT

The **Automatic Direction Finder (ADF)** is a medium frequency navigation aid that provides relative bearing information. The ADF system is a form of radio compass; that is, the ADF is a sensitive receiver that will "home in" on a specific radio station. By knowing the location of two or more stations, the pilot can determine aircraft heading and position, that is, establish a "fix."

In order to establish a heading, the pilot flies towards a known station location, and in order to establish position (fix), the pilot must select two (or more) different ground stations, and triangulate to find her position.

The ADF system operates by tuning one station, but uses two different types of antenna. Each antenna senses a different portion of the radio wave signal from the same ground station. The signals are then combined to create a complex electronic signal, which is interpreted by the system to provide direction information. The two antennas are known as the **loop antenna** and the **sense antenna.** The ADF receiver processes both sense and loop antenna signals, and provides the result as a bearing indication given with respect to the nose of the aircraft.

Advantages of the system are

- It has a long range.
- It can use any medium frequency ground station.
- It is an ideal navigation aid for use in small aircraft.
- It has light weight and low power.

PRINCIPLES OF ELECTROMAGNETIC WAVE PROPAGATION

Radio waves have two parts. In a vertically polarized transmitter antenna, the electrostatic **E field** is vertical, and the electromagnetic **H field** is horizontal. These orientations are called polarization. With other transmitters, various polarizations are possible depending upon the design of the transmitter antenna.

The best situation for ADF is a vertically polarized transmitter and a vertically polarized receiver antenna. The ADF has a sense antenna that is vertically polarized, and a loop antenna that is horizontally polarized.

RF waves propagate at the speed of light, which is 300×10^6 m per sec. They are similar to light or heat transfer. The RF spectrum in general is 10 kHz to 10 GHz. The ADF spectrum is 200 kHz to 1799 kHz.

RF is transferred by one of three paths:

- ground wave—this is the practical usage for ADF.
- skip/sky wave—E and H fields are distorted and are no longer valid.
- direct/space wave—Not used.

RF wave propagation is affected by:

- day/night
- sunspots
- seasons

Antenna length can be determined by using the equation:

$$\text{lambda (wavelength) } \lambda = (\text{velocity} \div \text{frequency}) \times (\text{submultiple antenna length})$$

Antennas operate the best when the physical antenna length equals wavelength. Antennas can be designed to have wavelength of half, or quarter, or some other submultiple of the wavelength. A quarter wavelength dipole antenna operates almost as well as a half or full wavelength antenna. Commonly, the aircraft skin forms one of the poles of the dipole antenna; this is known as a "ground plane." Most avionics receivers and transmitters use quarter dipole wavelength antennas. The ground plane will most often be the aircraft's skin. This can be a problem in newer composite material aircraft.

Figure 6.1 shows the reception pattern of a loop antenna (bidirectional or two lobed) and the reception pattern of a sense antenna (omni). These radiation patterns are a function of the design of the antenna and not of the polarizations. The ADF loop antenna is tuned to the electromagnetic H field. The ADF sense antenna is designed to pick up the electrostatic E field.

Figure 6.1
Antenna Radiation Characteristics

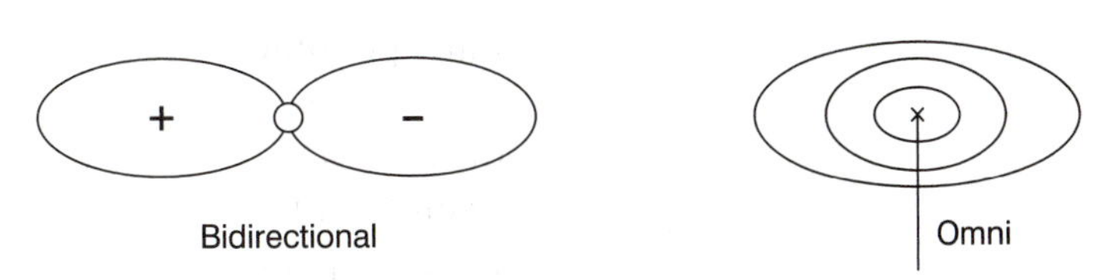

ADF THEORY

The ADF system operates in the **medium frequency (MF)** band from 200 to 1750 kHz. In this frequency band two different kinds of transmissions can be employed by ADF. They are **NDBs** and **NAVAIDs.** NDBs are nondirectional beacons such as AM commercial radio stations. NAVAIDs are specifically designed for ADF navigation. NAVAIDs will transmit a Morse code identifier. A pilot will usually select a high power radio station to provide her with a signal over long distances, and then when she is closer to her destination she will use the NAVAID.

There are three different styles of ADF design. When you read about ADF systems at a technical level you will find that the ADF system operates in a similar manner, but the electronics of the receiver and associated electronics is slightly different. In all cases the output feeds the indicator and provides relative bearing information.

The receiver installation consists of the following:

- receiver
- an azimuth indicator
- sense antenna
- loop antenna

The receiver provides controls of the audio level, test function, on/off, and frequency selection. There are two operational modes for ADF, conventional receiver (AM) and ADF mode. In some models continuous wave (CW) mode is available.

Loop Antenna

The loop antenna receives RF waves in a reception pattern similar to a figure eight (see Figure 6.2). The loop antenna, which is rectangular and can be mounted on the aircraft belly or roof, is a circular loop of coiled wire in a

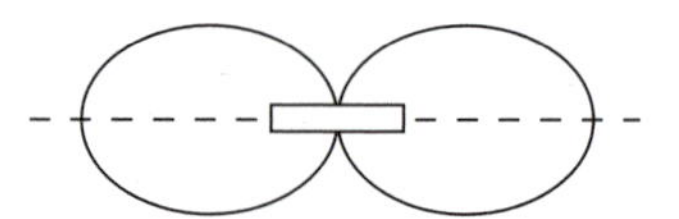

Figure 6.2
The Reception Pattern in the Plane of the Loop Antenna

fixed loop. There are two insulated coils of wire wound at right angles on a ferrite core. The maximum voltage is induced into a coil when the antenna coil is perpendicular to the transmitter. The loop antenna cannot distinguish whether the signal is from the 0° position or from the 180° position. The loop antenna uses the null position as it is finite compared to the wide angle of the maximum signal position.

In a fixed loop antenna system, *e*1 and *e*2 (Figure 6.3) are induced voltages from transmitted radio waves. These voltages reproduce the RF field within the body of the **goniometer** (shown in Figure 6.4). During the "direction finding" or DF operation, the goniometer rotor automatically rotates to obtain a signal null. At this position a minimum voltage is induced into the rotor. The azimuth indicator needle shows the relative bearing of the transmitting station.

Figure 6.3
Ferrite Core of ADF Antenna

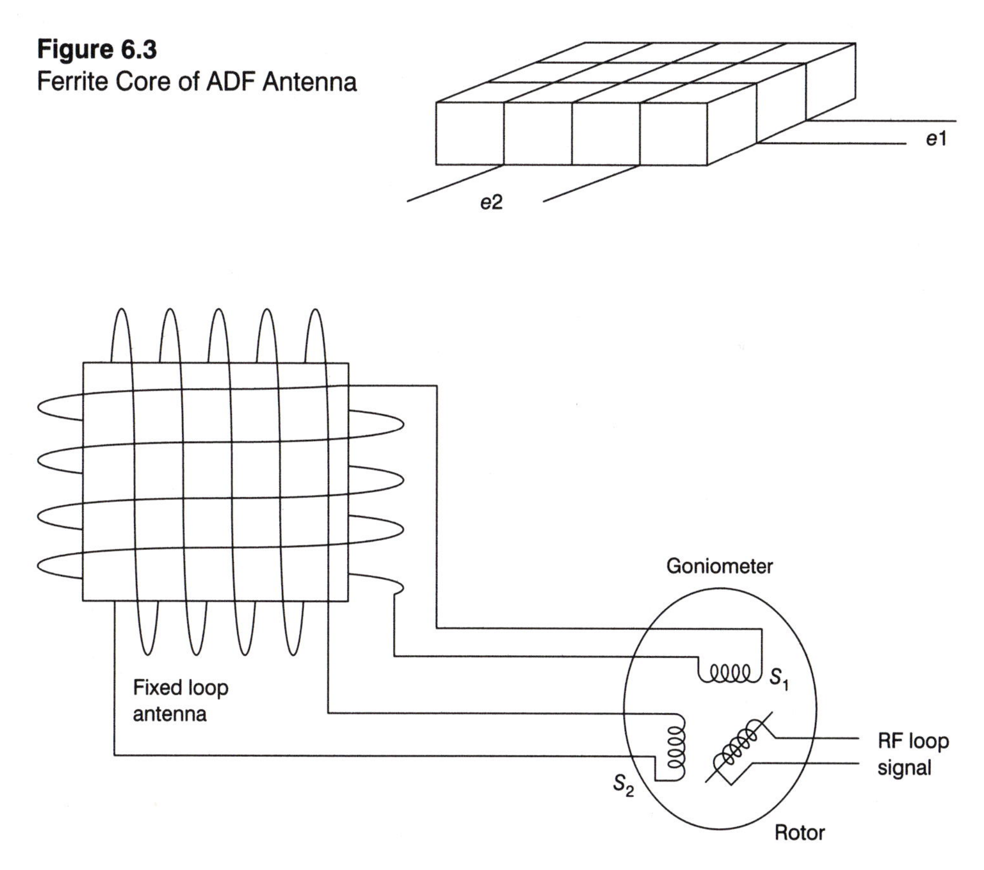

Figure 6.4
Electrical Outputs from ADF Fixed Loop Antenna

Sense Antenna

The sense antenna is either a vertical whip antenna (rare), or a long wire horizontal style. The sense antenna is a wire from 2 to 5 m long mounted from the cabin roof to the vertical tail fin. The antenna is mounted with insulating standoffs. It can also be a whip type like the one used on most car radios. The important parameter of a sense antenna is its effective height. The sense antenna receives its RF wave as a circular pattern, or more accurately, an **Omnidirectional** pattern.

Composite

The ADF receiver combines the two signals to create the equivalent of a **cardioid** pattern. In more modern type antennas, both loop and sense are in the same teardrop-shaped housing. All ADF antennas are mounted as near to the aircraft centerline as is practicable.

The resultant **null zone** is an area where minimum (zero) signal is received, as shown in Figure 6.5. As with many electronics applications, the null zone is easier to tune than the maximum zone. The null is much more discrete than the maximum zone and therefore can be positively tuned. The ADF system electronically and/or mechanically aligns the null with the transmitter station, and presents this information as a relative bearing that is referenced to the aircraft's nose.

The ADF receiver tunes in the signal from the selected transmitter at two points. They are:

- the bidirectional loop signal
- the Omnidirectional sense signal

Figure 6.5
Cardioid Pattern Showing the Null and the Maximum Zones

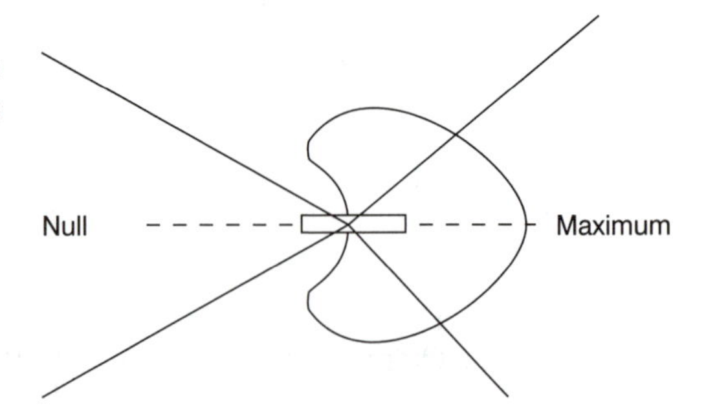

These two signal patterns are combined as a cardioid pattern with the appropriate polarities assigned. They provide the ADF with a reception pattern for determining the direction of the transmitted signal. The loop signal from the goniometer is not present when the system is nulled. If the goniometer rotor is not in the null, the loop signal will be developed. The loop antenna signal either leads or lags the sense antenna signal by 90° depending on which side of the null the rotor is positioned. For example, if it is detected that the loop signal lags the sense signal by 90° because the goniometer is clockwise of the null, then a signal is developed to drive the rotor counterclockwise. This will drive the rotor into the null zone. Once the null is reached, the drive signal disappears and the system is nulled. Opposite conditions also apply.

Sources of ADF Error

There are four sources of ADF error:

1. Vertical effect—caused by an unbalanced loop antenna circuit
2. Quadrantal error—caused by reradiation from the aircraft structure (wings, tail, props)
3. Night effect—caused by partial horizontal polarization of the incident waveform
4. Shore effect—caused by a change in direction of a wave crossing from land to water, or vice versa.

Block Diagram and Details

Figure 6.6 diagrams a typical closed loop ADF.

Because the fixed loop has no moving parts, it is nearly trouble free. It consists of two loops of wire oriented at 90° to each other. The voltage (both phase and magnitude) is fed to the two-stator winding of an RF resolver or goniometer, as shown in Figure 6.7.

If the station is directly in front of the aircraft, the *A* loop is at a maximum voltage and *B* loop is at a minimum or null (see Figure 6.8).

Figure 6.9 shows that the two signals from the fixed loop windings are combined by the stator of the goniometer and reproduce the magnetic field in the body of the goniometer. The rotor of the goniometer orients itself in such a manner as to find the null. The resolver rotor will see the null signal.

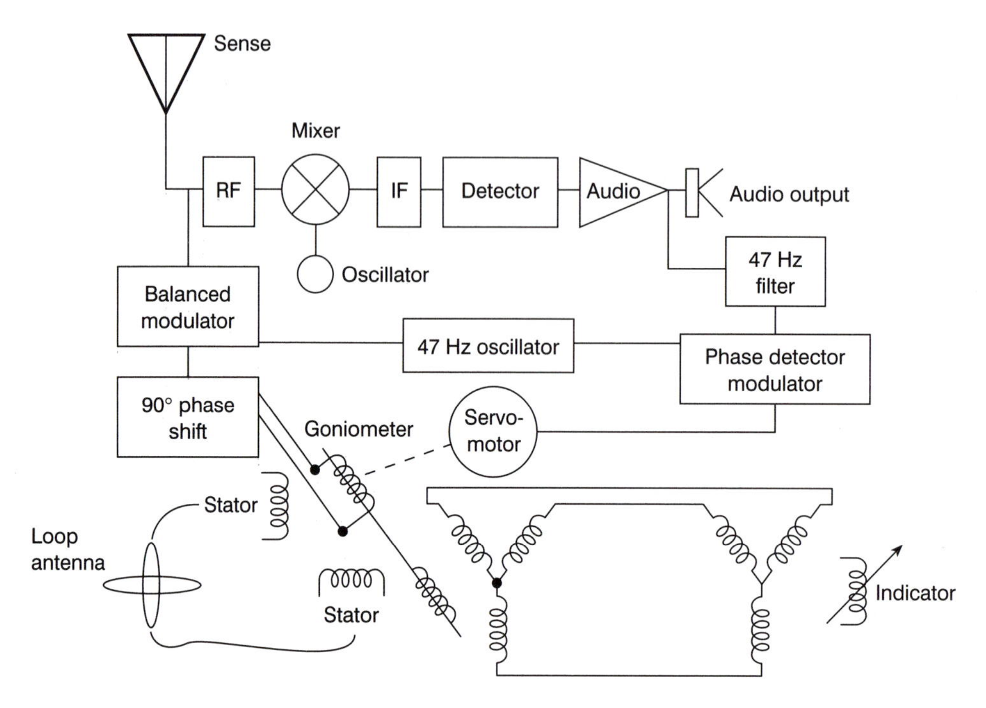

Figure 6.6
Typical Block Diagram of Closed Loop Automatic Direction Finder

Figure 6.7
Output Signals from Fixed Loop Antenna

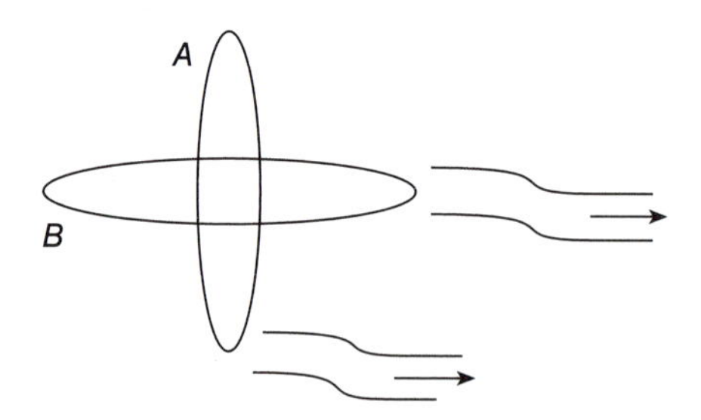

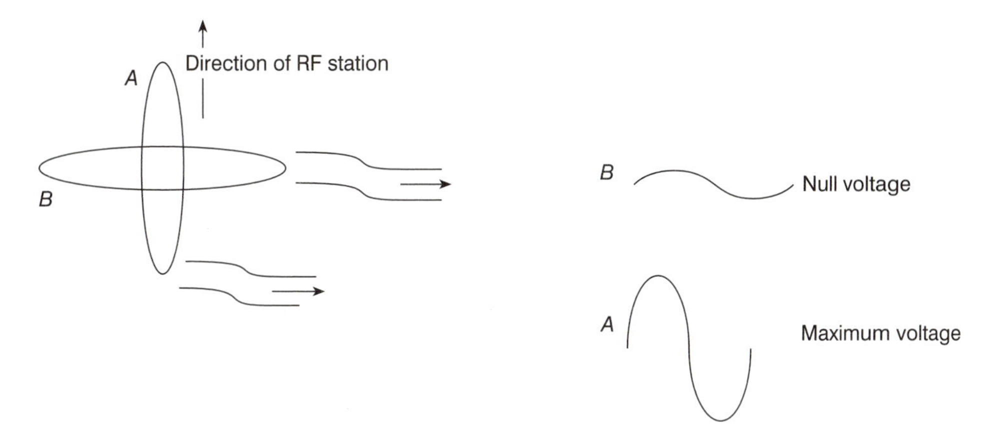

Figure 6.8
Relative Signal Strengths from Fixed Loop Antenna

Figure 6.9
Signal Pickup from Antenna to Receiver

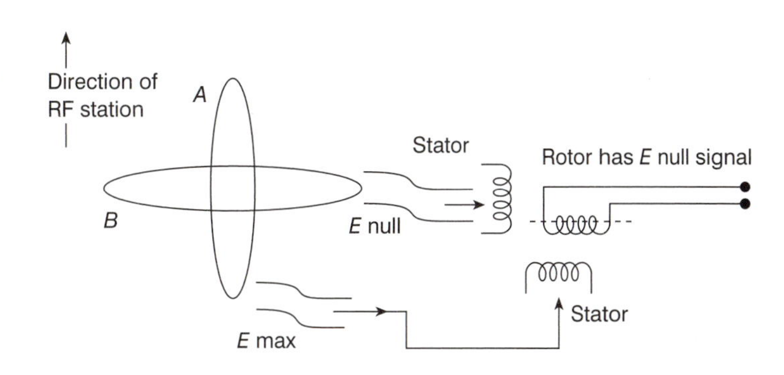

QUESTIONS

1. The ADF system is a very high frequency navigation system. (True or False?)
2. A cardioid pattern is obtained from the sense antenna. (True or False?)
3. The ADF system requires two separate antenna. (True or False?)
4. The pilot receives a signal that she can actually listen to, and also steer by. (True or False?)
5. The signals induced to the antenna(s) are phase shifted by 90. (True or False?)

6. The sense antenna has an Omnidirectional antenna pattern. (True or False?)
7. The ADF indicator shows the relative bearing of a transmitter. (True or False?)
8. The ADF antennas can be located anywhere on the aircraft. (True or False?)
9. ADF system frequency is from 200 to 1750 kHz. (True or False?)
10. ADF can have an audio Morse code station identification. (True or False?)

CHAPTER 7

Distance Measuring Equipment

DME

Introduction

This chapter deals with the basic principles of Distance Measuring Equipment.

Objectives

Sketch the basic transmission diagram for DME

Identify the operation frequency range of DME

Explain slant range error

Briefly explain pairing

State what data are usually presented to the flight crew

State the pulse spacing for X mode

State the pulse spacing for Y mode

Briefly explain search, track, and coast operations

Explain squitter

Explain the need for a suppression bus

Interpret the blocks of the block diagram

Basic Concept

Distance Measuring Equipment (DME) is a measurement system that provides the flight crew with distance information from the aircraft's position to a ground station.

DME CONCEPT

The term **Distance Measuring Equipment (DME)** means a measurement system that provides the flight crew with information about the distance from the aircraft's position to a ground station.

Greater accuracy, reliability and range can be achieved by using a system that does not rely on the reflection of energy, as in a radar system, but rather, relies on transmissions from both the aircraft and the ground station, as diagrammed in Figure 7.1.

The airborne DME (interrogator) includes:

- transmitter
- receiver
- timing circuits
- distance indicator

The ground-based DME (transponder) consists of:

- transmitter
- decoder/encoder computer—time delay
- receiver

The time of travel of the RF wave is measured from aircraft to ground, and back to the aircraft. The airborne unit subtracts the ground delay time of 50 μs, divides by 2, then displays the distance in NM.

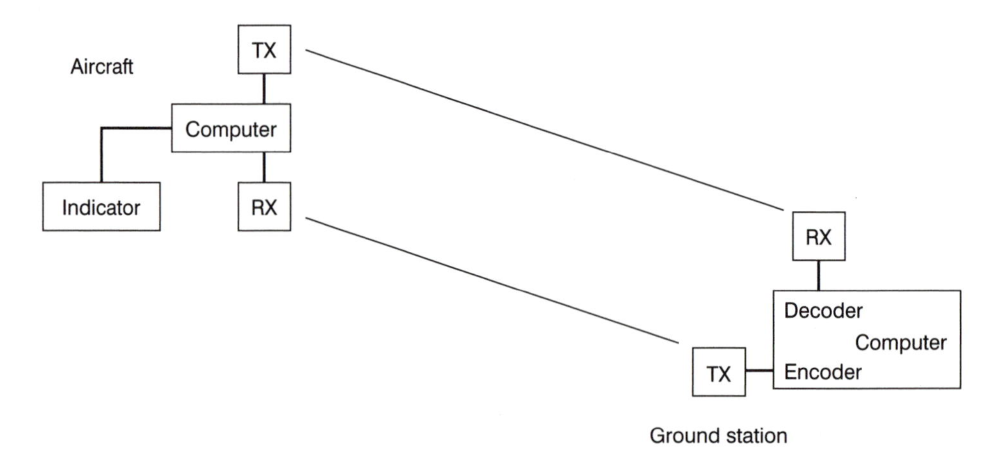

Figure 7.1
DME Transmission Diagram

Be careful with the use of the term *transponder.* **ATCTX** means the **air traffic control transponder** (an airborne unit) as it is used in an air traffic control situation, and the term **transponder** means a system in general that responds to interrogation signals. The ground station transmits on one frequency, and the aircraft DME transmits on a different frequency. DME operates in the L band between 960 and 1215 MHz. The DME is used primarily in conjunction with VOR/ILS equipment. When pilot selection is VOR stations, the DME can provide distance information up to 250 NM, depending on the unit. When selecting ILS frequencies, the DME is generally used within about 50 NM of the ground station. It should be noted that the practical range of a DME depends upon the aircraft's altitude, due to the line of sight operation of L band radio signals. The limited range due to altitude is caused by the curvature of the earth. A good rule of thumb to approximate the distance to the horizon is to calculate the square root of the aircraft's altitude above terrain, then multiply this by 1.25. The solution to this is the range to the horizon in NM. For example, with an aircraft at 10,000 ft above terrain, the DME would not be expected to operate at a range more than 125 NM.

Slant range is the actual line of sight distance to the station as determined by an electromagnetic wave, as shown in Figure 7.2.

Slant range error is the difference between the slant range distance and the actual ground distance. When the aircraft is at a low altitude (Figure 7.3) or is at a great distance from the ground station (VORTAC), the slant range is almost the same as the ground distance. Therefore the slant range error is small, as shown in example 1.

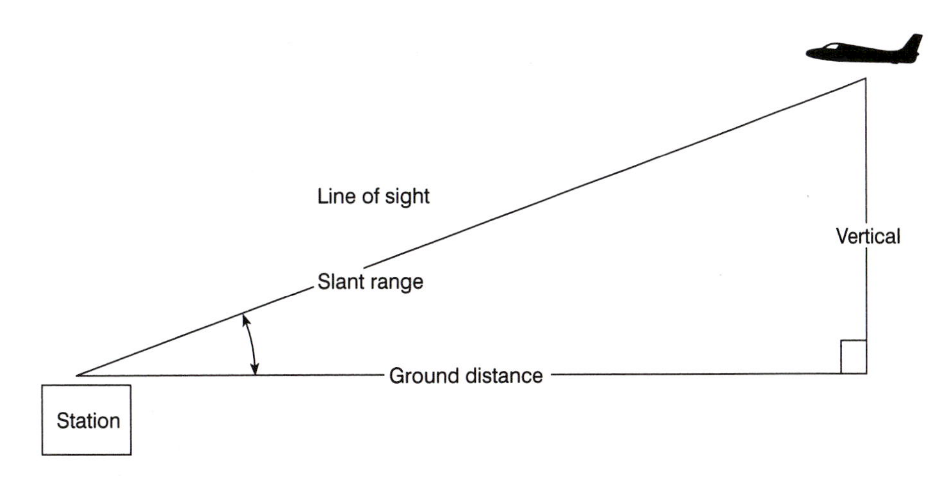

Figure 7.2
Slant Range

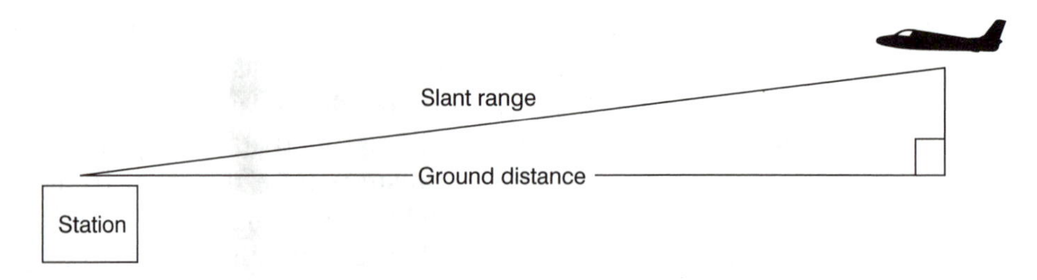

Figure 7.3
Slant Range—Low Altitude

Slant Range Variable	Example 1	Example 2
Altitude	20,000 ft	20,000 ft
Actual ground distance	50 NM	1 NM
DME reading	50.1 NM	3.5 NM
Slant range error	.1 NM	2.5 NM

DME presents the slant range. When the aircraft is closer to the ground station, especially at higher altitudes, the slant range and the ground distance become significantly different. In this situation the slant range error is large, as in example 2.

Aircraft position or fix can be obtained through a number of methods, either through Rho Rho or through Rho Theta navigation principles. In order to obtain the distance information required for navigation we use the DME. This information can be presented to the pilot for his use in navigation solutions, or the information can be electronically distributed to components and systems that use the information for navigation purposes.

PILOT'S PERSPECTIVE

The DME indicator can display either **ground speed (GS),** or **Time to Station (TTS),** depending on the pilot-selected controls. The frequency selection of the VOR or ILS also controls the frequency of operation of the DME system. In selecting VOR or ILS, the pilot initiates a lot of electronic circuitry, but sees only the presentation as DME information.

The pilot sees a DME presentation on a separate indicator like the one shown in Figure 7.4, although older units are analog.

He could see the same information presented on an HSI—Horizontal Situation Indicator, like the one shown in Figure 7.5.

Figure 7.4
DME Digital Indicator

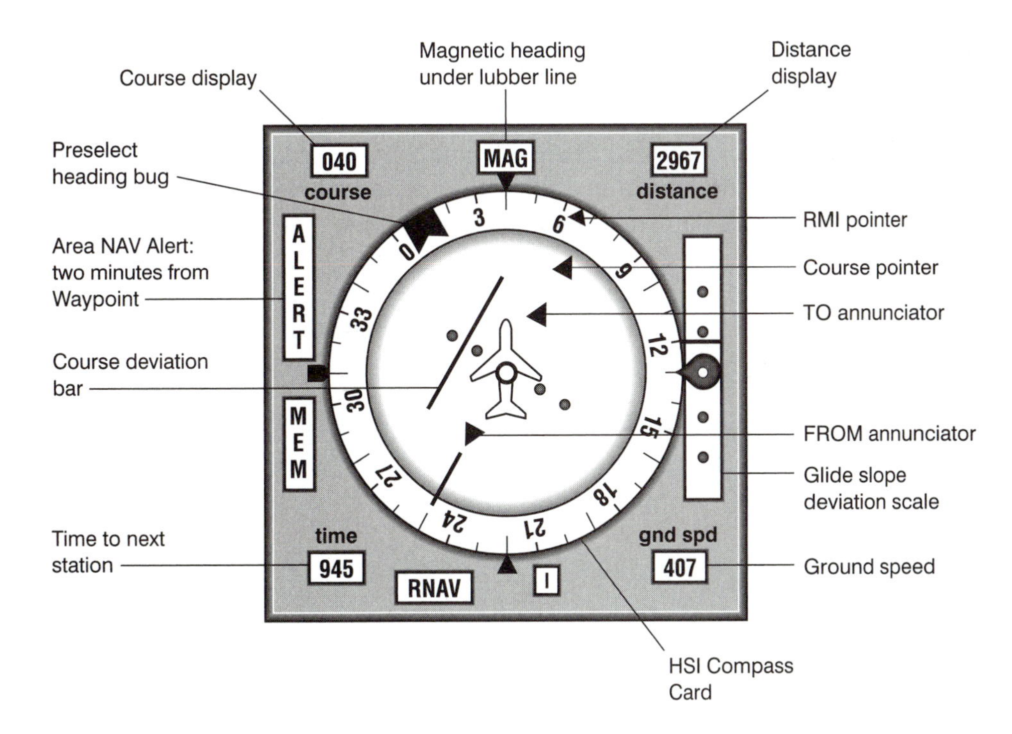

Figure 7.5
HSI showing DME functions

When an aircraft has two VHF NAV receivers, it is possible to select NAV 1 or NAV 2 and, using a double pole single throw switch, to operate the DME from either NAV receiver channel selection. DME is channelled by the VHF navigation head automatically.

For example, when a VOR station operates at 117.8 MHz, if the DME ground receiver operates at 1149 MHz, then the DME ground transmitter operates at 1212 MHz. In other words, DME transmitter and receiver frequencies are always 63 MHz different from each other regardless of the station selected.

ELECTRONICS OPERATION

The DME system is actually the distance or range portion of the military Tactical Air Navigation (**TACAN**). When the TACAN and the VOR station are located at the same location, the station is referred to as a **VORTAC.**

There is a standard relationship between the VHF NAV frequencies and the DME frequencies. In other words, each DME frequency has a paired VOR or ILS frequency. When the pilot selects a frequency on his VHF NAV control head, the DME is automatically channelled to the paired frequency.

When the DME computer recognizes a reply to its interrogations, the system locks up and DME information is displayed. As indicated in Figure 7.6, the airborne transmitter then transmits two pulses. This pulse pair is transmitted at random spacing. The time spacing between the transmitted

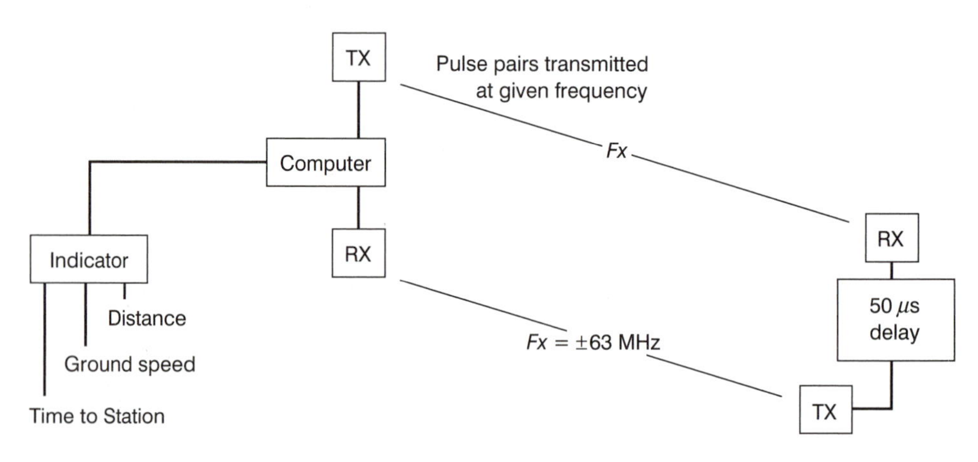

Figure 7.6
DME Transmission—Detail

pulses $P1$ and $P2$ is fixed, but the pulse pair repetition period is random. When this pulse pair is received by the ground station (interrogation), the ground station adds a 50 μs delay time, and then transmits a pulse pair (reply). The reply is transmitted at 63 MHz above or below the aircraft-transmitted frequency.

The ground station periodically ignores interrogations and transmits evenly spaced pulses that appear as Morse code at 1350 Hz to provide for station identification.

DME slant range is calculated in the DME computer by measuring the elapsed time between the interrogation pulse pair and the reply pulse pair. From this result, 50 μs is subtracted to compensate for the 50 μs fixed time delay inserted by the DME ground station. The difference is then divided by 12.36 (time duration of a radar mile). This results in slant range in NM. For example, the interrogation pulse pair and the reply pulse pair are measured to be 1286.0 μs apart.

$$1286.0 - 50 = 1236.0$$
$$1236.0 \div 12.36 = 100 \text{ NM}$$

Encoding is the exact spacing between two pulses in order to ensure that signals are valid between the aircraft and the ground station. The DME operates on two frequency bands, a high and a low band. There are two modes of operation—the X mode and the Y mode. The X mode has 12 μs spacing for both air-to-ground and ground-to-air transmissions. The Y mode is 36 μs for the air-to-ground mode, and 30 μs for the ground-to-air transmission. The ground station transmits 2700 pulse pairs per second regardless of the number of aircraft that are interrogating it. If there are not enough aircraft in the area using the DME to make up 2700 pulse pair returns, then the DME station inserts randomly spaced pulse pairs to make up the difference. From the point of view of an airborne receiver, the signal received is made up of one pair of pulses for each of that unit's interrogations (the reply pulses) and random appearing pulse pairs. These random pulses are called **squitter.**

As the aircraft is maneuvered, the aircraft can shield the line of sight transmission and cause loss of signal. In the airborne circuitry a memory holds the last calculated data for about 20 sec before the loss of signal is reported to the display indicator. If the signal is lost then the DME will "coast" for 10 sec before the circuitry decides that it has lost tracking; it then initiates the search and lock-on procedures. In search mode there is a high pulse repetition frequency of approximately 115 **PRF** looking for the correct reply. The warning flag shows DME *not* operating.

In track mode there is a low pulse repetition frequency of approximately 25 PRF indicating DME has found the correct reply, and the DME flag is

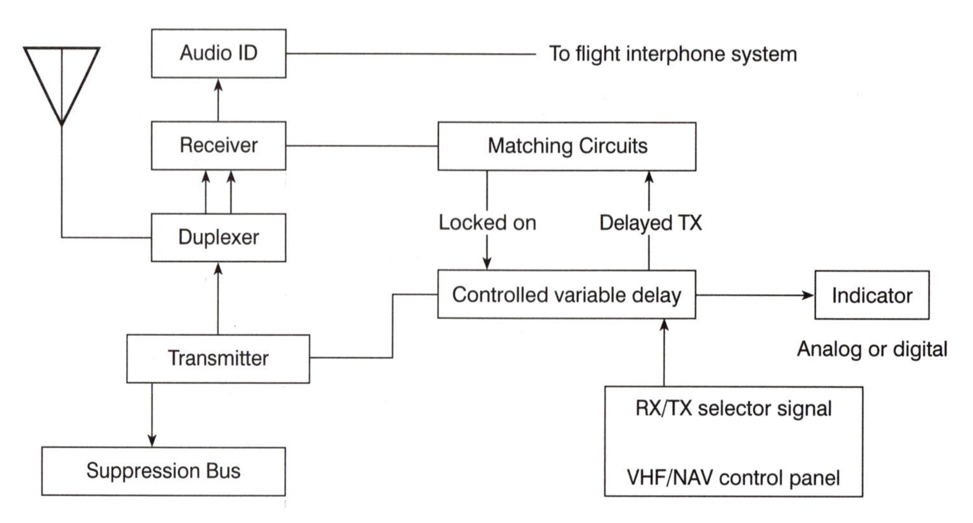

Figure 7.7
Typical DME block diagram

detented. This provides the ground station opportunity to service an increased number of aircraft.

The hold function allows the DME to retain the information about the last frequency that it was dialed to, even when the NAV has been rechannelled.

DME and the transponder (ATCTX) operate on similar frequencies. In order to prevent any interaction between transponder and DME, three things must take place:

1. Antenna separation—the DME antenna must be located at least 6 ft away from the transponder antenna,
2. Suppression of the DME when transponder is transmitting, and
3. Suppression of the transponder when DME is transmitting.

The purpose of the **Suppression Bus** (see Figure 7.7) is to suppress the DME when the transponder is transmitting, and to suppress the transponder when the DME is transmitting. DME and the transponder will interfere with each other if allowed to do so, resulting in errors or possible damage to one or both systems.

The ATCTX frequencies lie in the center of the DME frequency spectrum. As the DME is tuned to 112.30 from 112.20, the DME system must "jump over" the transponder frequencies.

Antennas must be located on the bottom side of the aircraft and as far apart as is practicable.

QUESTIONS

1. The DME is usually used in conjunction with:
   a. VOR
   b. ADF
   c. Loran C
   d. GPS
2. The airborne DME is called:
   a. transmitter
   b. interrogator
   c. receiver
   d. UHF pulse pair
3. The DME system requires at least _________ transmitters.
   a. one
   b. two
   c. three
   d. none
4. What is the time duration for a radar mile?
5. What is the pulse spacing of channel X airborne transmitter?
6. What is the pulse spacing of channel X ground transmitter?
7. What is the pulse spacing of channel Y airborne transmitter?
8. What is the pulse spacing of channel Y ground transmitter?
9. Does DME have an audio signal?
10. Explain "squitter."
11. Explain when the slant range error is greatest (range and altitude).
12. Explain when the slant range error is least (range and altitude).
13. How many pulse pairs does the ground station transmit per second?
14. Explain the difference between DME search mode and DME track mode.
15. What is the ground station transmission delay time?
16. Explain DME hold.
17. Explain the purpose of the suppression bus.
18. Explain coast mode.
19. How is the DME channelled?

CHAPTER 8

Area Navigation

RNAV

Introduction

This chapter introduces the concept of Area Navigation with some insights into operation, design, installation, and maintenance requirements.

Objectives

Provide a basic area Navigation Solution

Solve geometric problems by using the sine law and the cosine law

List the required navigation signals to effect proper RNAV operation

Define the terms "Waypoint offset" and "Waypoint bearing"

Describe the pilot inputs to RNAV

List the modes of operation

Define the term "constant course width"

Describe RNAV system operation when an ILS frequency is selected

Sketch RNAV system block diagram

Basic Concept

Area Navigation (RNAV) is a guidance system that uses DME slant range, VOR bearing, and pilot inputs to compute bearing and distance to a Waypoint. The pilot flies to the Waypoint as though it were a real VORTAC station.

AREA NAVIGATION CONCEPT

In the traditional methods of navigation, the pilot selects a series of VOR/DME stations that provide enroute information. The track is a series of zigzags from station to station through to the destination. In the **Area Navigation (RNAV)** system, the RNAV computer can electronically change signals from these ground stations so they appear to be relocated. If the pilot relocates the stations to be in a direct route between source and destination, the trip can be shortened significantly.

VOR and VORTAC stations are usually identified on aeronautical charts by showing a Compass Rose surrounding the station, identifying the VOR with its radio frequency, and giving the three-letter Morse code identifier. See Figure 8.1.

If the VOR is colocated with a DME, then it is often referred to as a **VOR/DME** and can be shown as a six-sided figure inside a square, as seen in Figure 8.2.

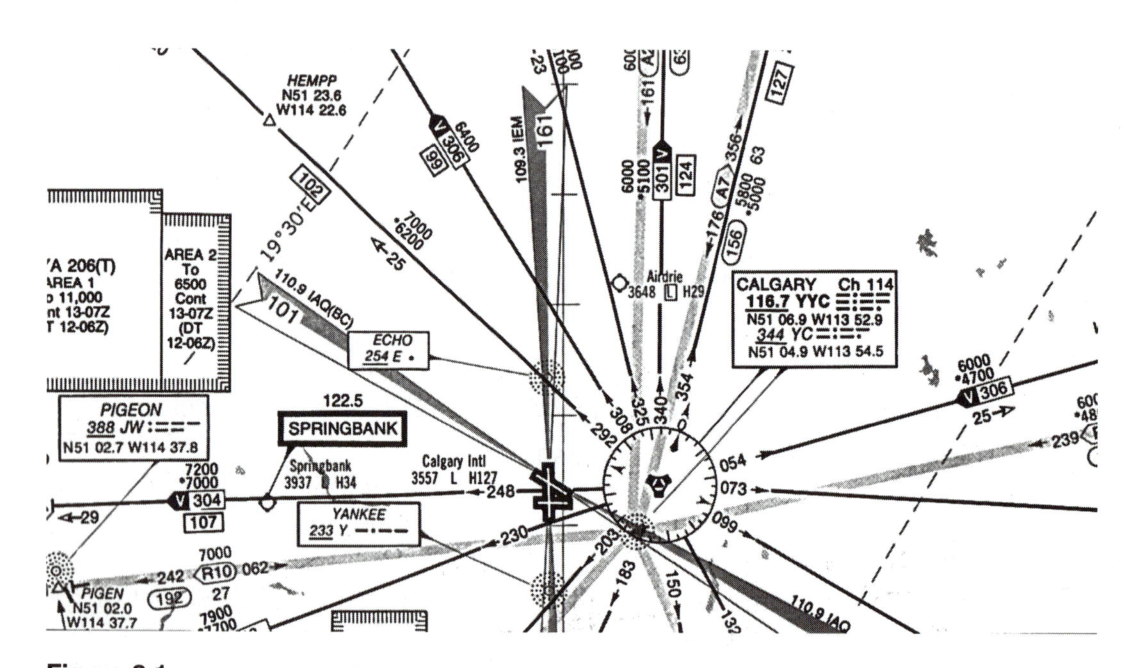

Figure 8.1
VOR Station Identifier
Source: © 1996 Her Majesty The Queen in Right of Canada. Used with permission from the Department of Natural Resources (Geomatics Canada).

Figure 8.2
VOR/DME Map Symbol

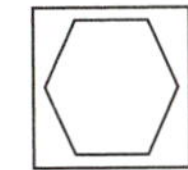

WHAT THE PILOT SEES

The pilot sets the VOR navigation receiver to the frequency as indicated on the aeronautical chart. She then verifies that she has the correct VOR by listening to the Morse code identifier. By centering the needle on the CDI, she can determine the bearing to the station. By "flying the needle," that is, keeping the needle centered, she will fly over the selected VOR. The DME will tell the plane's distance to the VORTAC.

To use RNAV, do the following:

- Calculate the desired position of the Pseudo Station.
- Set controls on the RNAV unit for bearing and distance from the VOR to the Pseudo Station.
- Select RNAV mode.
- Fly to the Pseudo Station using CDI and DME indicator as if the Pseudo Station were an actual VOR/DME.

The system that is driving the CDI or HSI must be annunciated to the pilot. This is usually accomplished with a set of lights, for example, RNAV, **Loran, Omega, INS,** etc.

RNAV CAPABILITIES

With the aid of the RNAV computer, the flight crew can electronically position VORs in such a location that there appears to be a VOR in more or less a straight line from origin to destination. This method of navigation greatly reduces the actual distance travelled, and saves cost and time. RNAV can reduce airway congestion by allowing the pilot to select a route that is parallel to an existing airway. It can also provide for more precise holding patterns, procedure turns, and runway threshold monitoring as well as simplified **STOL (Short Takeoff or Landing)** and helicopter navigation solutions. RNAV allows for **Standard Instrument Departures (SIDS),** and **Standard Terminal Arrival Routes (STARS).**

When we electronically position a station, the new station is referred to as a **Waypoint** or a **Pseudo Station.** The pilot can set the distance and bearing to the Pseudo Station from the VOR/DME or VORTAC.

"Waypoint offset" is a pilot input; it is the distance that the Pseudo Station is to be electronically moved.

"Waypoint bearing" is a pilot input; it is the direction that the Pseudo Station is to be electronically moved.

Electronically the station can be positioned anywhere, but more practically, the RNAV computer requires actual information from real VOR/DME or VORTAC stations and must be in reception range of these stations. The approximate range of VORTAC is about 150 NM, but this distance does depend upon the aircraft's altitude. On aeronautical charts some Pseudo Stations are actually named, and the charts come complete with both Waypoint bearing and Waypoint offset. A four-point star is used as the map designation for the Pseudo Station, as shown in Figure 8.3.

Here is a bird's-eye view of the problem. We know the distance and the bearing to the VOR. We wish to electronically "move" the station to a new location, then fly directly to it.

The concept is further explained in the following paragraphs.

Standard VOR Route Example

Figure 8.4 shows the course for an aircraft flying from Calgary to Cold Lake using VOR. The aircraft leaves the Calgary airport and takes a direct route

Figure 8.3
Area Navigation Problem

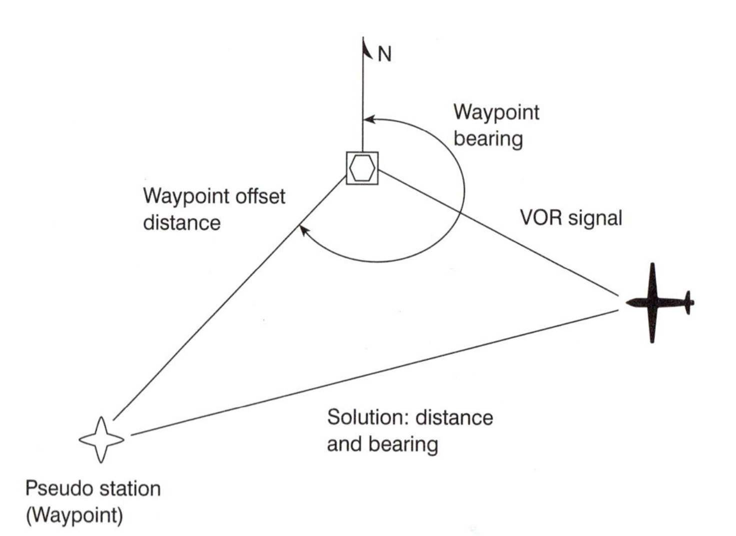

Figure 8.4
VOR Navigation Solution

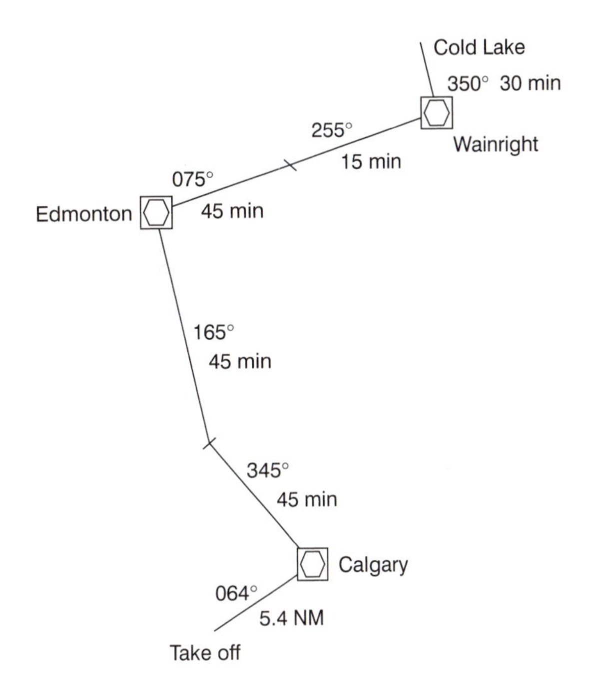

to the Calgary VOR, 064° @ 5.4 NM (DME). The aircraft turns left on the 345° radial outbound. About 45 min later the Edmonton VOR is used, and the aircraft catches the 165° radial inbound. After 45 min the aircraft passes over the Edmonton VOR and takes a right turn on the 075° radial outbound. After another 45 min the Wainwright VOR is used, and the aircraft now flies inbound on the 255° radial. Within 15 min the aircraft passes over the VOR and turns left on the 350° radial outbound. Approximately 30 min later the pilot can start looking for the Cold Lake airport.

RNAV Route Example

Figure 8.5 shows the much more direct course made possible by RNAV. The pilot electronically moves the Edmonton and Wainwright VORs into a direct line with the Cold Lake VOR. (Waypoint 1 is moved in the direction of 120° with a distance of 90 NM and Waypoint 2 is moved in the direction of 300° with a distance of 60 NM.) This makes the course considerably shorter, resulting in a direct flight from Calgary to Cold Lake.

The distance and the direction of the new Pseudo Stations must be programmed into the RNAV computer by the pilot. Note that during each leg of the flight, the pilot must reselect the appropriate VOR frequency. In other words, during the center leg, the pilot still uses the Edmonton VOR/DME, but does not need to fly over Edmonton.

Figure 8.5
Area Navigation Solution

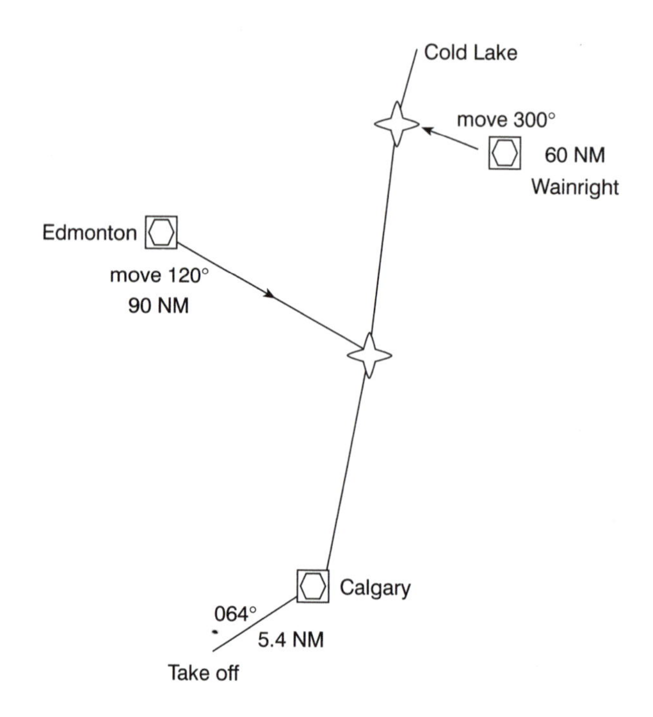

MODES OF OPERATION PILOT PERSPECTIVE

The advantages of RNAV in navigation solutions lie in the many ways in which the flight crew can use this navigation system. There are four distinct modes of operation.

VOR/DME Mode

In the VOR/DME mode, the RNAV computer performs the tasks of a VOR converter circuit. In normal VOR navigation solutions, the course deviation indicator's course deviation bar indicates to the pilot the angular error that the aircraft is displaced away from the selected course. The actual distance from the selected course depends on the distance from the station. The more that the D Bar is displaced from the center, the greater the angular error. This provides a wedge-shaped corridor that the aircraft must stay within as it flies toward the station, as seen in Figure 8.6.

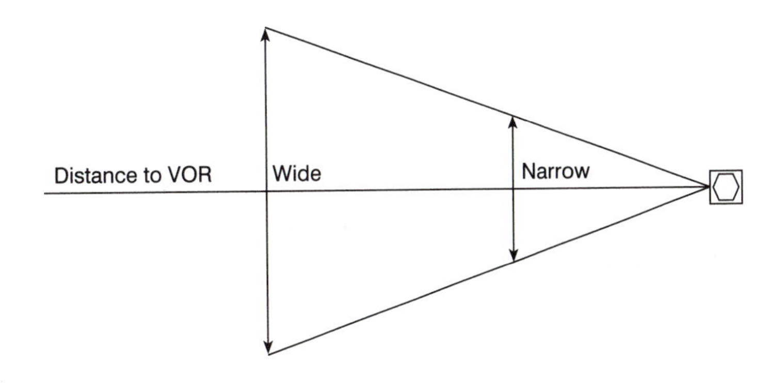

Figure 8.6
Wedge-shaped VOR Deviation

Enroute or RNAV Mode

Enroute or RNAV mode calculates the distance and bearing to a Waypoint. This mode provides a constant course width, not the traditional VOR wedge-shaped pattern, as shown in Figure 8.7. The course is 10 NM wide. If the CDI shows full-scale deflection, the pilot knows that the aircraft is 5 NM from the centerline of the selected course.

A half-scale deviation on the CDI indicates to the pilot that the plane is 2.5 NM from the centerline of the selected course. Two advantages of the constant course width (in comparison to the standard VOR wedge-shaped course) are (1) a holding pattern shaped like a symmetical racetrack (as in Figure 8.8) instead of the wedge-shaped holding pattern of VOR and (2) a smoother station passage. The aircraft does not fly through the Cone of Silence.

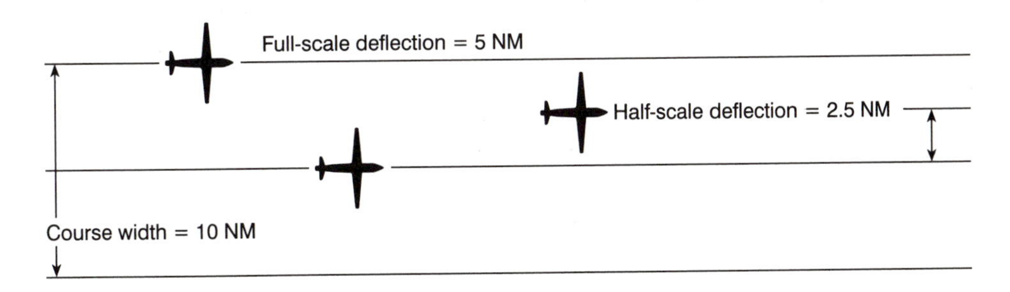

Figure 8.7
CDI in Enroute or RNAV Mode

Figure 8.8
VOR (left) and RNAV (right) Holding Patterns

Approach Mode

Approach mode is more precise than the Enroute mode, because it reduces the course width to 2.5 NM, as shown in Figure 8.9. In RNAV, the deviation bar (D Bar) represents the distance from the selected course. Selecting the approach mode changes the sensitivity of the D Bar from 5 NM to 1.25 NM for full-scale deflection.

Parallel VOR

Parallel VOR is like Enroute mode, except that the Waypoint offset is set to zero. This provides for a constant course width instead of the converging width of standard VOR. This option may not be present in all RNAV systems.

RNAV COMPUTERS

The RNAV computer often uses a microprocessor. The solution to the establishment of Pseudo Station locations is accomplished by the RNAV computer. If the computer is analog, then the solution uses the electronic addition of input sinusoidal or square wave signals. If the computer is digital, then the navigation solution uses the sine and cosine laws.

Figure 8.9
CDI in RNAV Approach Mode

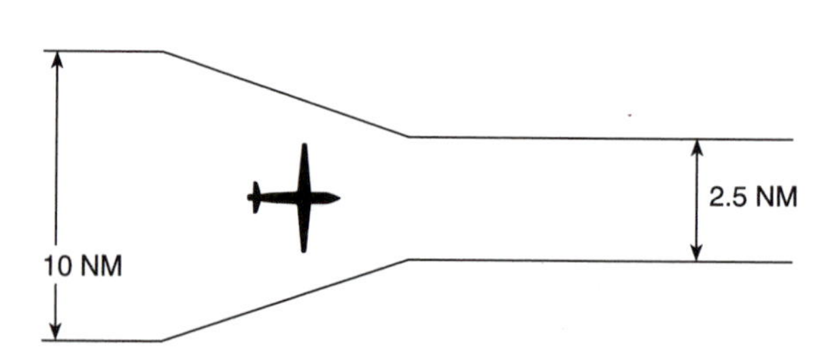

The Digital Computer

A digital logic computer may have a database and is capable of computing a variety of navigation solutions. The computer uses inputs from the VOR system, the DME system, and the pilot. These inputs need to be converted to the computer's language. Then the mathematical calculations can be performed by the computer. After the computer has completed the equations, the digital signal must again be converted to the correct format for use by the indicators. Some RNAVs are capable of **frequency agility,** meaning that they can process information from more than one VORTAC simultaneously.

In order to understand what the RNAV computer is doing, it will be helpful to perform a mathematical calculation. Given the triangle in Figure 8.10, find the length of side *A* and the angles *b* and *c.* The solution requires the use of the sine and the cosine laws. Capital letters *A, B,* and *C* are the sides of the triangle, and *a, b,* and *c* are its angles. Note that side *A* is opposite angle *a,* etc.

In any right angle triangle, the sine value of the angle is the ratio between the length of the opposite side and the length of the hypotenuse. The cosine value of the angle is the ratio between the length of the adjacent side and the length of the hypotenuse, and the tangent value of the angle is the ratio between the length of the opposite side and the length of the adjacent side.

The *sine law* identifies the relationship between the angles and sides of a triangle.

$$\frac{A}{\sin a} = \frac{B}{\sin b} = \frac{C}{\sin c}$$

Figure 8.10
Triangle Navigation Solutions

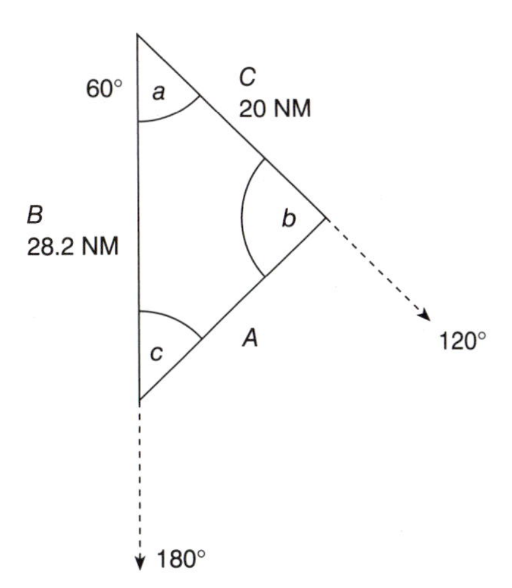

The *cosine law* allows for the calculation of an unknown side of a triangle when the other two sides and one angle are known.

$$A = \sqrt{(B^2 + C^2 - 2BC\cos a)}$$

Referring to Figure 8.10, using the cosine law, first solve for the length of side A.

$$A = \sqrt{28.22 + 202 - 2(28.2 \times 20) \times \cos 60}$$

$$A = 25.1 \text{ NM}$$

Then using the sine law, solve for the two unknown angles, b and c.

$$\sin b = \frac{B \sin a}{A}$$

$$\sin b = \frac{28.2 \sin 60}{25.1} \qquad b = 76.65°$$

$$\sin c = \frac{20 \sin 60}{25.1} \qquad \text{c} = 43.6°$$

Here is a typical example of the calculation that the RNAV computer performs in order to provide the navigational solution. As shown in Figure 8.11, an aircraft is located on the 120° radial and 53 NM from the VOR.

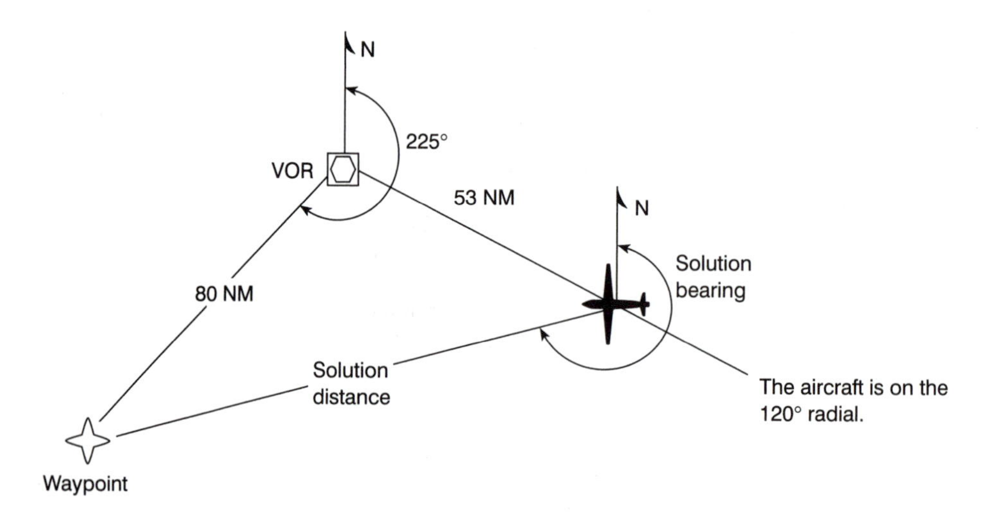

Figure 8.11
RNAV Triangle Solution

The pilot wishes to fly to a Pseudo Station. The pilot input to the RNAV will be:

Waypoint offset, 80 NM, and
Waypoint bearing, 225°

The RNAV computer then must calculate both the solution distance and the radial from the Waypoint on which the omni bearing selector (OBS) will center the CDI. In this example the OBS will center on 254°, and the Waypoint will be 107 NM DME. The computer must then perform the calculation, and the technician must understand how it was accomplished if he wishes to repair an out-of-service system.

Figure 8.12 shows a diagram of a typical RNAV system. *Note:* Most general aviation RNAVs don't have a central air data computer input.

Input Signals

The Area Navigation computer requires signals from the VOR receiver and the DME receiver. The VOR receiver provides angular information, and the DME system provides distance information. The installation technician must be careful to ensure that the correct signal is getting to the RNAV units, as some DMEs will supply information as digital and others will supply information as analog.

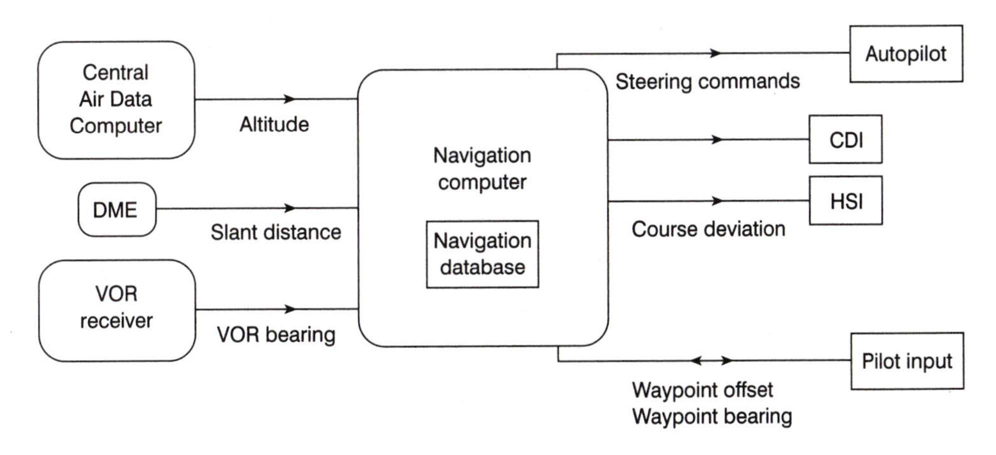

Figure 8.12
Block Diagram of a Typical Area Navigation System

To be more specific, the information from the NAV receivers is VOR radial information—a VHF demodulated signal from the VOR receiver. The signal from the DME system may exist in one of many formats.

The DME analog distance on some units is as small as 40 mV/NM. In the case of a digital DME, a clock, sync, and data pulse, usually conforming to ARINC standards, are used. Another option is a pair of pulses, the time separation of which is proportional to the DME distance.

System Interaction

Proper operation of the RNAV system depends on the reception of the DME and VOR signals from the same location. If the NAV receiver were rechannelled and an ILS frequency selected, then the signals would be inappropriate for accurate Waypoint display, and the RNAV operation would stop. Normal VOR/DME operation would resume. Also if the DME were placed on hold, and the NAV rechannelled to a different VOR, then the DME portion of the RNAV solution would be inaccurate. Usually the RNAV enroute or approach mode will disable the DME hold function.

The Analog Computer

The input signal from the pilot is a sinusoid signal with an input voltage amplitude proportional to the distance that the Waypoint is to be located from the VORTAC. The phase angle of this signal is dependent on the Waypoint bearing (pilot input). The input from the VOR system and the DME system is combined to produce a second sinusoidal waveform, the phase angle of which is dependent upon VOR bearing, and the amplitude of which is dependent upon DME distance. These two signals will vectorially add and the resultant will be the solution to the Waypoint's location. This resultant signal will then be appropriately processed and sent to the CDI or HSI to provide the deviation bar with course activity.

Some RNAV analog processors use square waves much like what was just described for the analog computer.

Miscellaneous Notes

The majority of the newer and updated RNAV units are able to preprogram a number of Pseudo Stations (Waypoints) so that RNAV inflight reprogramming is not necessary.

Figure 8.13
Quick Solution to Right Angle Triangle

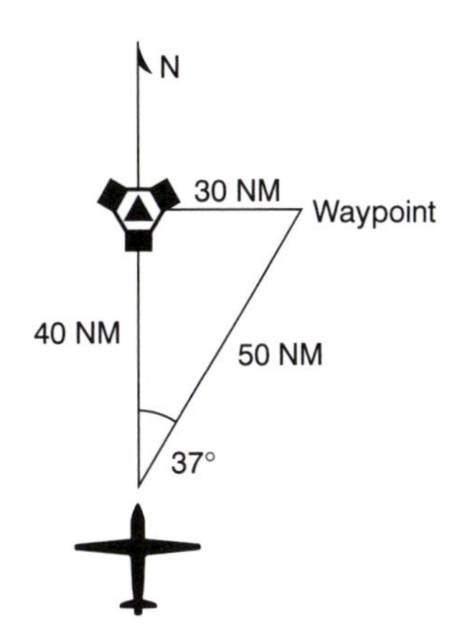

An additional use of the RNAV is to relocate the off airport VOR/DME to a runway location of a non-ILS approach. The RNAV is now a useable approach aid.

There is an easy method for the technician to verify that RNAV is doing the sine and cosine solutions correctly. This is the old carpenter's trick of the three-four-five triangle. Using the facts given in Figure 8.13,

set the Waypoint bearing to 090°

set the aircraft on 000°

set the DME to 40 NM

The solution is

DME reads 50 NM

OBS centers at 37° "to"

Generally the best way to install RNAV is with a three-unit package from one manufacturer—the DME, the VOR, and the RNAV computer. It is possible to mix and match different manufacturers' products, but this requires interfacing boards, and may not work as well as you would like.

QUESTIONS

1. Define Area Navigation.
2. Explain how VOR/DME navigation is accomplished. Sketch and label a diagram.
3. What are the advantages of using RNAV?

4. Define Pseudo Station or Waypoint.
5. Can a Waypoint be positioned anywhere? Explain.
6. Can a Waypoint be used when an aircraft is out of the VORTAC signal range?
7. What are the two purposes of RNAV?
8. Sketch a block diagram of a typical RNAV system. Show inputs and outputs.
9. A normal CDI D Bar indicates the angle between the aircraft heading and the selected course. This angle remains constant but the distance of the aircraft to the selected course gets smaller. Explain the operation of the D Bar in the RNAV mode.
10. What input variables are needed in order to interface the RNAV computer with the autopilot?
11. Can a Waypoint be fixed in a three-dimensional position?

CHAPTER 9

Weather Radar

WX

Introduction

This chapter provides an introduction to the concepts of weather radar with some insights into design, installation, and maintenance requirements.

Objectives

Explain the general principles of radar

Explain the term "echo"

Explain radar mile

Describe typical function controls for the radarscope

Sketch and label the system block diagram

Basic Concept

Weather radar provides weather information to the flight deck.

WEATHER RADAR

Radar is an acronym for ***RA*****dio *D*****etection *A*****nd *R*****anging.** It is a high-power RF energy wave transmitted outward, and the energy that returns is used to determine the density and distance of the target from the radar transmitter. It is generally accepted that RF energy propagates at 6.2 μs/NM using space wave propagation. A radar mile is then considered to be the time it takes for the wave to go to the target, and then return—12.36 μs. Radar systems typically function in the SHF band from 3 to 30 GHz.

The information is most useful to the flight crew as range and bearing information. These systems can often provide terrain mapping. The returned energy reflected by the target is called the *echo.* The energy return, dependent upon the size of the target, can be caused by aircraft, weather, storm cells, terrain, etc. Range of the radar is limited by line of sight, rather than by power. For example, an aircraft at 39,000 ft would see the horizon at 247 NM due to the curvature of the earth.

In C band frequencies, wavelength is 7 cm, whereas in X band, wavelength (lambda) is 3 cm. As the frequency increases with a small antenna we can detect only large targets. This is because, for higher frequencies, a small high gain antenna is practical.

The bearing is determined by the angle between the antenna and the aircraft nose. Normally the top of the CRT indicates the heading of the aircraft. The range can be adjusted from between 2.5 and 250 NM (for some models). This information is presented on a cathode ray tube (CRT).

Modes of Operation

Weather radar (WX) has several modes of operation:

Standby mode—in warm up mode for the electron tube

On—actual operational mode: transmit and receive

Test—shows color bars and ranging, used for calibration

Weather/NAV—provides information from the navigation computer, such as distance and course to the next Waypoint

EFIS—some electronic flight instrument systems allow radar information to be presented on the CRT

Tilt—allows the flight crew to manually adjust the pitch position of the antenna

Range select—allows the operator to select the range of the CRT display

Target alert—indicates that a strong return is being received ahead, but beyond the selected range on the CRT

Weather alert—puts a flashing condition on the magenta colored areas to flag them as critical to the flight plan

Weather conditions are identified by color on the display CRT:

Black areas indicate clear weather conditions.

Green areas indicate light clouds.

Yellow areas indicate increasing thickness of the weather condition.

Red indicates a concentrated weather formation.

Magenta indicates an extreme weather condition (such as hail or thunderstorm) or terrain (cumulo granite—very solid weather).

Microcomputer-based radar systems provide for self-monitoring of the internal workings of the system, and allow the maintenance technician to evaluate and adjust color, accuracy of the transmitted beam, and many other parameters.

The system block diagram in Figure 9.1 shows a simplified block diagram of a typical weather radar. The antenna transmits RX energy to the target and captures the echo energy. The TX and RX are synchronized so that a high energy pulse is *not* returned to the receiver. The computer controls timing and graphics generation, and the display has some pilot controls as listed previously under modes of operation. Also see the discussion of target energy early in the next chapter.

Figure 9.1
Typical Block Diagram of Simplified Weather Radar

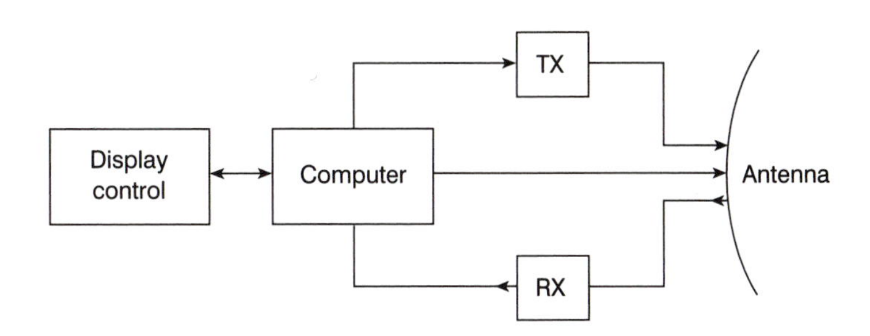

QUESTIONS

1. Explain the concept of radar.
2. What is the frequency range of radar?
3. What is the time duration of a radar mile?
4. What is a CRT?
5. How does the radarscope display increasing severity in the weather?
6. Explain the tilt function.
7. What is the typical weather display range?

CHAPTER 10

Transponder

ATCTX

Introduction

This chapter deals with the principles of the Air Traffic Control Radar Beacon System and Air Traffic Control Transponder.

Objectives

Identify Air Traffic Control Radar Beacon System and its function

Identify ATC and its function

Identify and explain function of Air Traffic Control Transponder

Explain the PSR basics

Explain SSR basics

Identify the transponder and antenna

Explain optimum antenna location

Explain how the pilot uses the transponder

State the frequency of the ground transmitter

State the frequency of the airborne transmitter

State interrogation modes of operation and describe each mode

State the reply modes of operation and describe each mode

Analyze the block diagram

Basic Concept

The transponder is the air-to-ground link in the Air Traffic Control Radar Beacon System.

The **Air Traffic Control Transponder (ATCTX)** is the airborne unit that responds to the ground interrogation. Ground interrogation comes from the ground radar. The transponder provides coded information to **Air Traffic Control (ATC)** about aircraft altitude and identity. The transponder is often abbreviated as ATCTX or XPNDR. It is the airborne part of the **Air Traffic Control Radar Beacon System (ATCRBS).** This system is an aid to navigation, as the transponder provides the Air Traffic Control System with information about the aircraft, and in turn, this information is useful for the safe routing of the aircraft by the Air Traffic Controller. Remember that all moving aircraft in the control zone are under the authority of Air Traffic Control.

AIR TRAFFIC CONTROL SYSTEM

The Air Traffic Control System provides a service for expeditious aircraft movement through control zones and specified airways. A **Control Zone** is the space in which the ATC is given the responsibility for all air traffic. This space is approximately 10 mi in diameter and 10,000 ft high, usually in proximity to an airport. A specified airway is an airspace that is specifically identified as being under the control of the enroute Air Traffic Controller. Remember that the pilot is ultimately responsible for the aircraft's safety. The ATC System allows for the tracking of many aircraft, and generally provides the Air Traffic Controller with aircraft type, identification, registration, altitude, range, bearing, and airspeed data. ATC gathers and processes this data from communications with the aircraft and from its own radar system. ATC has two types of radar that are used to provide the air traffic controller with information.

The first type of radar is called the **Primary Surveillance Radar (PSR).** This is standard echo or reflection radar. (See the chapter on weather radar.) Briefly, when we think about normal radar transmission, the total target or returned energy is:

$$\text{Target energy} = \frac{K \text{ energy transmitted}}{\text{distance}^2}$$

$$\text{Echo energy} = \frac{K' \; (K \text{ energy transmitted})}{\left(\frac{D^2}{D^2}\right)}$$

where K' is a factor of the target size

where D is the distance to the target

where K is a constant for any given situation.

$$\text{therefore: echo energy is } = \frac{KK' \text{ (energy transmitted)}}{D^4}$$

This is sometimes referred to as the double distance square loss rule. This means that a normal radar pulse must travel out from the transmitter, and then bounce off of the target (skin return), and then return to the receiver antenna. In doing all of this, the signal is returned at a level that is extremely small compared to the original transmission. This of course limits the distance and quality of the signal, and therefore the quality of the system.

The disadvantages of normal radar are:

- The transmitter power must be high.
- There are different skin returns from different sizes and shapes of aircraft.
- Due to the width of the beam, the accuracy is lower than with a narrow beam.
- There is leakage from standard parabolic antennas, called side lobes.
- Stationary objects such as buildings are seen as targets (fixed ground returns).
- Weather will interfere (clutter) the radar display.

The transponder overcomes many of the problems associated with echo radar.

The second type of radar is the **Secondary Surveillance Radar (SSR).** This system operates on the *coded reply from the airborne transponder.*

The information from both PSR and SSR is used to provide the air traffic controller with a total air traffic situation, with information displayed on a single radar scope. The purpose of the transponder is to facilitate Air Traffic Control. Primarily it reinforces the signal on the radar scope, and provides positive identification of a radar PIP (target). This has the effect of extending radar coverage, and enables ATC controllers to more effectively control high-density traffic. In 1992 there were 900,000 scheduled flights in the United States.

TRANSPONDER

Air Traffic Control Transponder is an airborne transmitter and receiver that transmits in response to a ground interrogation. The ground radar site sends off a routine SSR interrogation. The SSR antenna is a smaller antenna, usually mounted on top of the PSR antenna. The aircraft transponder recognizes this routine interrogation, and then returns a signal to the ground station, which shows up on the Air Traffic Controller's radar console. The transponder's function is to transmit a coded response to a coded interrogation.

The advantages of the SSR system are:

- It has more range than a PSR for the same power.
- The return is the same size for any aircraft. The transponder return will be about 20 dB greater than a normal skin return. This equates to a power level of about 100 times greater than normal! More details of the mathematics involved are in Appendix B.
- It has no ground clutter.

The disadvantages of the SSR system are:

- It can produce **fruiting;** that is, when the ATC Transponder is being interrogated by more than one ground station, replies to one ground station are received by the other.
- It can produce **garbling,** or overlapping of the replies from two aircraft from the same interrogator.

The ATC ground station transmits at 1030 MHz. This is called the interrogation. The transponder must first recognize the signal; then it has to reply. Reply frequency is 1090 MHz. These frequencies are in what is commonly known as the **L Band** of the frequency spectrum. The DME operates in the same frequency band as the transponder, so care must be taken to prevent interference between these two systems. Physical antenna separation is often enough, but when this is not possible, a mutual suppression line connects the transponder to the DME. The suppression line causes an inhibit function of the transponder or DME decoder circuit whenever the other systems transmitter is in operation. To avoid damage to either DME or transponder, whenever one is transmitting the other is disabled. An ATC Transponder is a legal requirement for aircraft flying at greater than 18 000 feet (FL 18). As of 1992, all aircraft flying into or from controlled airspace must also be equipped with a transponder.

WHAT THE PILOT SEES

The pilot on the preflight "walk around" can see the transponder antenna located on the bottom of the aircraft. This antenna is about 3.5 in. in both height and length. It is often located toward the nose of the aircraft, and as far removed from the DME antenna as possible.

Once inside the cockpit the pilot may see a transponder like the one shown in Figure 10.1. There is a wide selection of manufacturers.

Figure 10.1
ATC Transponder (courtesy of Allied Signal General Aviation Avionics)

The transponder's control functions are:

- *On/Off/Standby:* This maintains the electronic circuitry ready to transmit, but alleviates the problem of ground station over loading due to high-density traffic. The pilot puts the transponder into standby mode when requested to do so by ATC. Standby mode also provides time to allow the transmitter electron tube to warm up. Without this "automatic turn on standby" time, the unit may transmit while cold and cause damage to the tube. A 1994 **NOTAM** (Notice to Airman) recommended leaving ATCTX in the "on" position when changing codes. The old procedure was to switch to standby, change code, then switch to "on." Unfortunately there was a tendency to forget to turn transponder back on, hence the NOTAM.
- *The "ident" push button:* When requested to do so by ATC (**Squawk Ident**), the pilot presses the ident button and a special code is transmitted to the ground station. This provides ATC with a positive method of identifying the aircraft on the ground radarscope; a double flashing parallel line indicates the aircraft's position for about 20 sec.
- *The mode select switch,* which allows the pilot to select several modes:

  Mode A: for standard domestic operation

  Mode B: for European operation

  Mode C: for the altitude transmission to ATC (see Appendix D)

  Mode A/C: for both Mode A reply and Mode C reply
- *The code selection,* which sets the coding for the transponder reply.

The Octal 4-digit number allows the pilot to communicate with the ground station, and the coding provides the ground station with various kinds of conditions or emergency situations that could exist on board. There are 4096 possible codes.

One of the conditions could be the aircraft flying a VFR flight plan under 18000 feet; this would likely be a code 12XX. Another example would be a code 14XX, indicating that the aircraft is an IFR flight below 18000 feet.

A general emergency response is automatically initiated to ATC when the transponder code selector is set with a 7 as the first digit.

For example, a setting of 7X00 would elicit emergency responses at ATC. The second digit identifies a more specific type of emergency:

7700: onboard critical emergency (Mayday)

7500: a hijacking

7600: a radio failure (In aviation slang this is referred to as **NORDO**—no radio!)

Naturally ATC will respond in a different fashion to any given emergency.

- *The reply light:* Each time the ground interrogator interrogates the transponder and each time the transponder replies, the reply light will flash. It's a very small light, and not distracting to the pilot, but does confirm that the system is operational.
- *The hi/lo sensitivity switch:* When close to a ground interrogator, this switch reduces the sensitivity of the transponder (older units).

GROUND INTERROGATION COMPOSITE SIGNAL

Figure 10.2 shows the composite signal from the ground radar. The carrier frequency of the ground radar transmission is 1030 MHz. The diagram depicts signal strength versus azimuth. Pulses $P1$ and $P3$ are radiated from one antenna, and $P2$ is radiated from a different omnidirectional antenna. The side lobes are unwanted radiated signals from the antenna. This composite signal is not from the PSR, but rather from the SSR.

GROUND INTERROGATION CODING CHARACTERISTICS AND MODES OF OPERATION

The pulse amplitude and sequence interrogation from the ground antenna looks like Figure 10.3.

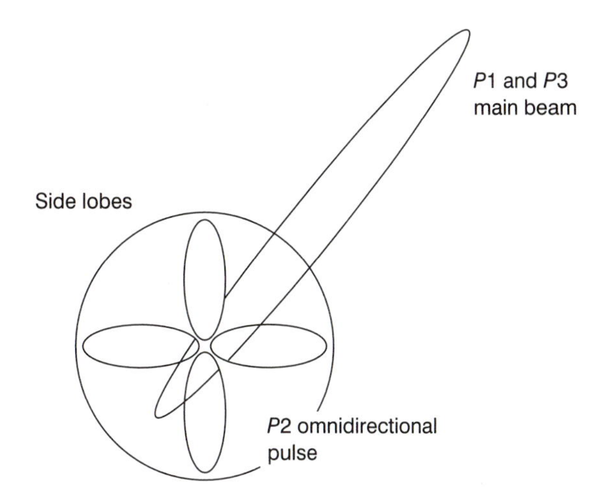

Figure 10.2
Composite Signal from the Ground Radar

All the pulses have the same pulse width of 0.8 μs. These pulses are created by transmitting the 1030 MHz carrier. The time from *P*1 to *P*2 is 2 μs, and the time from *P*1 to *P*3 determines the mode of interrogation. Note that the amplitude of *P*2 is smaller than that of both *P*1 and *P*3. The ground station interrogation rate is approximately 400 times per second. The purpose of the *P*2 omnidirectional pulse is to allow the transponder to determine if the interrogation came from the main beam or from a side lobe interrogation of the SSR. If interrogated by a side lobe, the transponder will not reply. In a **side lobe interrogation,** the *P*2 pulse is greater than or equal to the *P*1 pulse. Comparison of amplitudes is performed by a circuit called a "ditch digger." In a main lobe interrogation the *P*1 pulse has more power than the *P*2 pulse. The difference must be 9 dB before the signal is guaranteed to be accepted by the transponder as a valid **main beam interrogation** (main lobe ground interrogation). See Figure 10.4.

The SSR interrogates the transponder in one of three modes. The type of interrogation is determined by the pulse spacing of the SSR carrier. Mode *A* is the common ATC interrogation mode, and Mode *C* is for pressure altitude. Mode *B* is used in Europe and Mode *D* is unassigned at this time. The

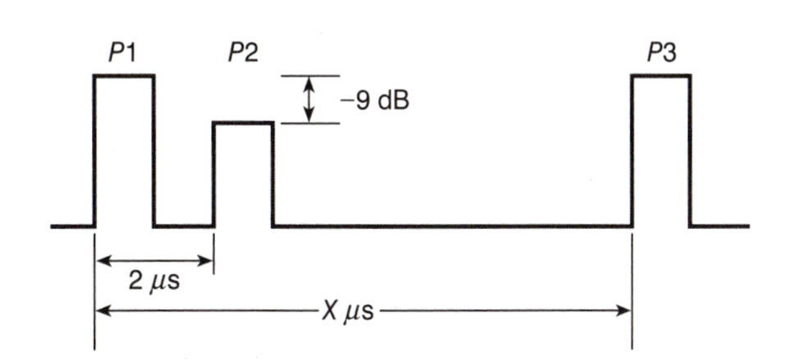

Figure 10.3
Ground Interrogation and Coding Characteristics

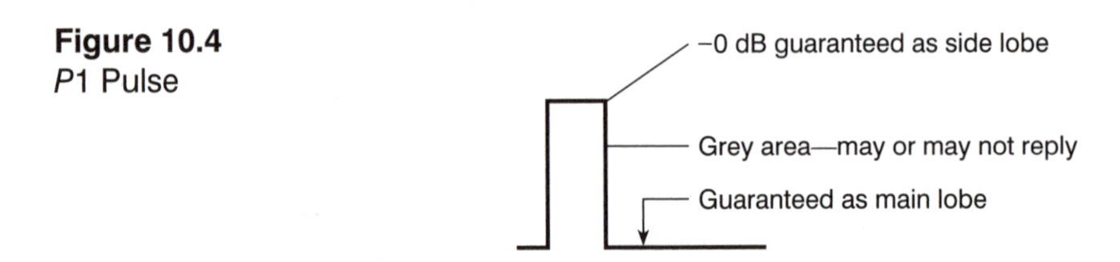

Figure 10.4
*P*1 Pulse

modes are established by the time between *P*1 and *P*3 pulses. Pulse spacing is Mode $A = 8\ \mu s$, Mode $B = 17\ \mu s$, Mode $C = 21\ \mu s$, Mode $D = 27\ \mu s$. Figure 10.5 shows Mode *A* and Mode *C* with time and amplitude characteristics.

REPLY FORMAT

After being interrogated, the airborne transponder returns a reply based on the setting of the code selector switches, as set by the pilot. The format is in the form of digital pulses having pulse widths of 0.45 μs with 1 μs spaces between the pulses. The pulses are surrounded by two framing pulses referred to as *F*1 and *F*2. This means there is a position for a reply pulse

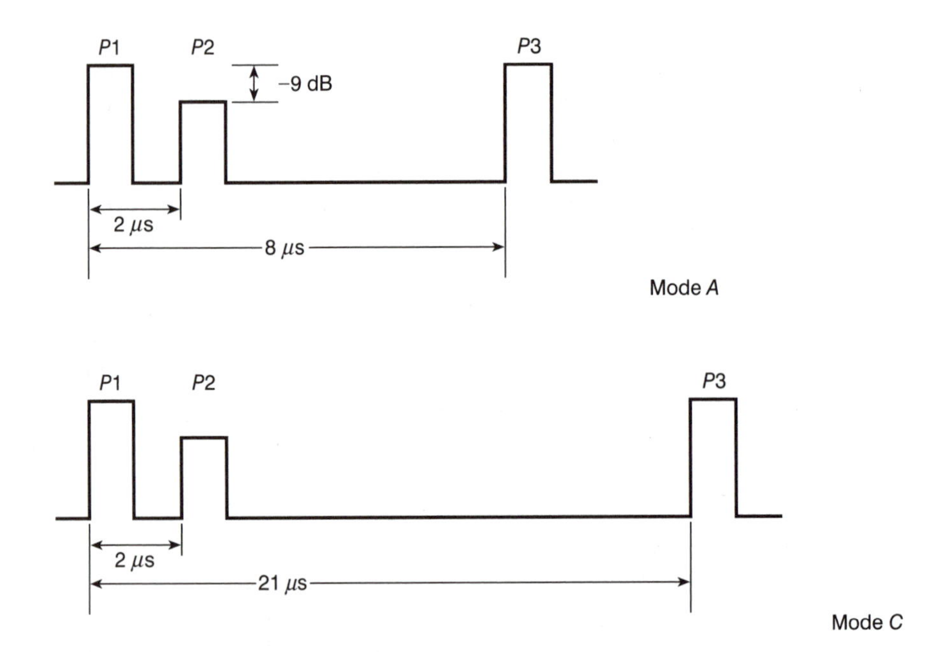

Figure 10.5
Transponder Modes of Operation Timing Characteristics

Figure 10.6
Spacing of Reply Pulses

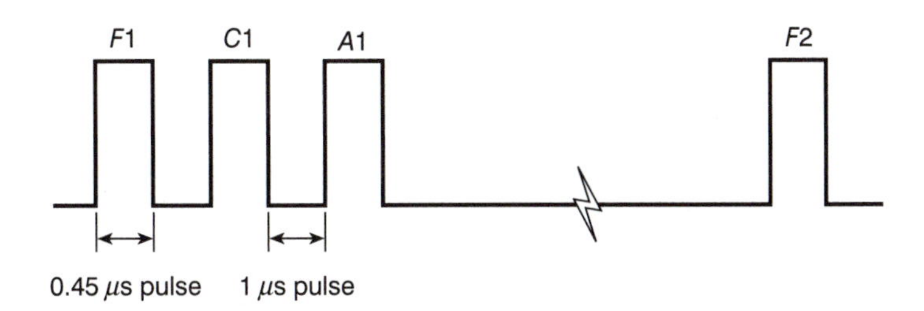

every 1.45 μs between $F1$ and $F2$, as shown in Figure 10.6. The pulses are coded, then transmitted in the following sequence:

$$F1, C1, A1, C2, A2, C4, A4, X, B1, D1, B2, D2, B4, D4, F2$$

The data string is enclosed between $F1$ and $F2$. They are coded this way in order to correspond with the Octal code; this allows electronic conversion from Octal to Binary. Note that the X pulse does not actually exist; it is just a space in the pulse string. For example, a pilot-set code of 1257 is really an Octal number. The Octal numbers are from 0 to 7. The Octal number 1 is placed in the A positions, 2 is in the B positions, 5 is in the C positions, and 7 in the D positions. For the A Octal number, a 1 is transmitted with a weighted value so the pulse $A1$ would be transmitted. The reply format for a code of 1000 is illustrated in Figure 10.7.

The pulse string has the $B2$ pulse for the Octal number 2. The reply code for 0200 is shown in Figure 10.8.

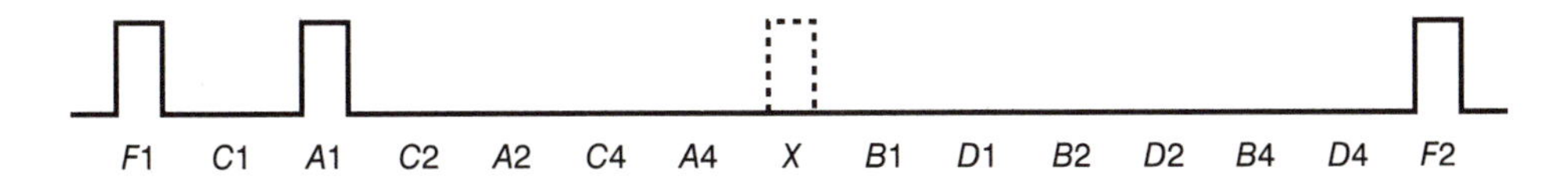

Figure 10.7
Reply Pulse Format

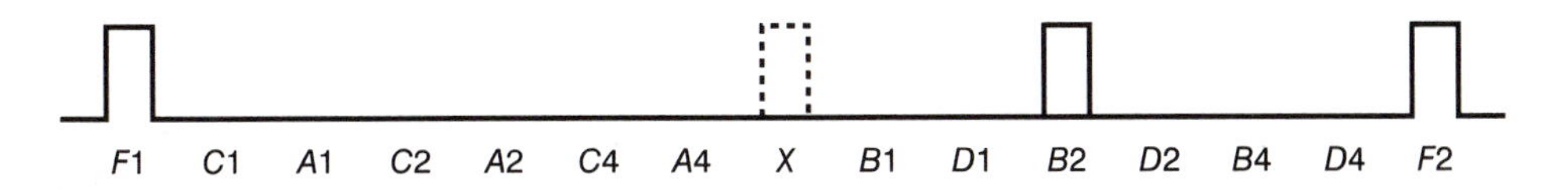

Figure 10.8
Reply Code for 0200

The pulse code for *C* digit of 5 is the combination of a 4-weighted pulse and a 1-weighted pulse; thus the reply format for a code of 0040 is as shown in Figure 10.9.

The pulse code for a *D* digit of 7 is the combination of a 4-weighted pulse, and a 2, and a 1; thus a reply format for a code of 0007 is as shown in Figure 10.10.

The reply code for a 1247 looks like Figure 10.11.

The emergency code of 7777 would have all of the pulses in the string, and the code would look like Figure 10.12.

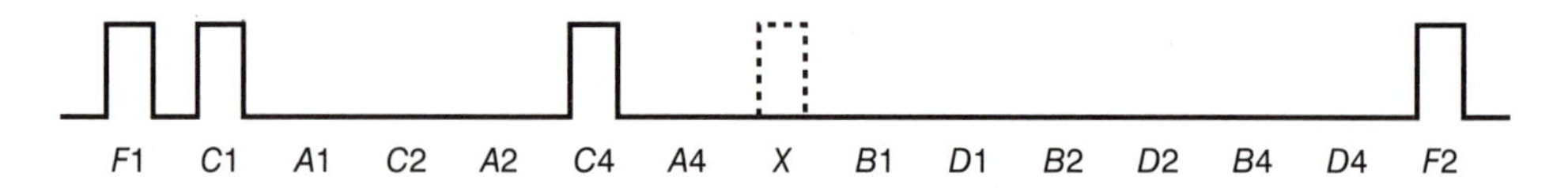

Figure 10.9
Reply Code for 0040

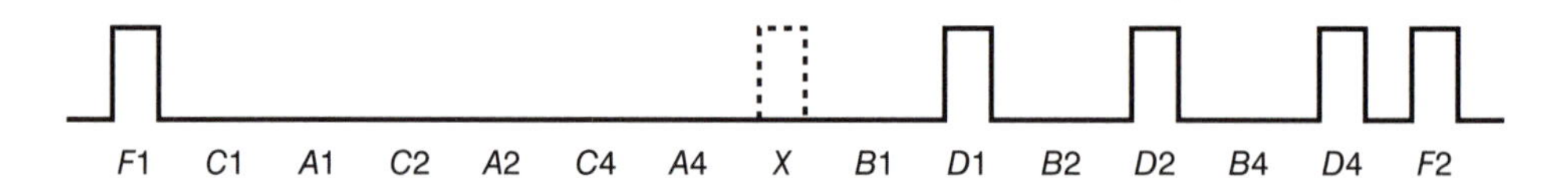

Figure 10.10
Reply Code for 0007

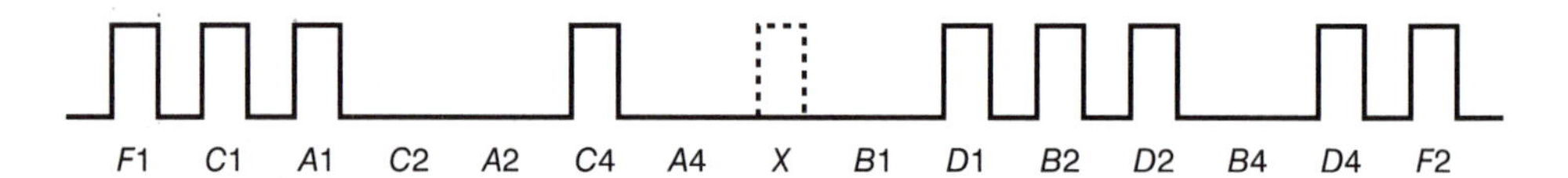

Figure 10.11
Reply Code for 1247

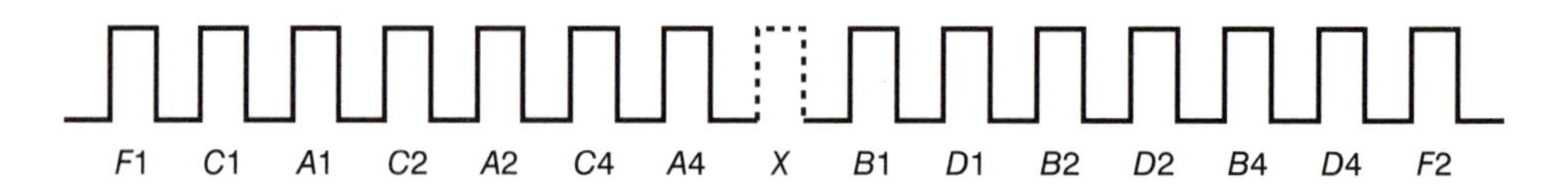

Figure 10.12
Reply Code for 7777

SPECIAL PULSE IDENTIFICATION FOR IDENT FEATURE

The special pulse identification, sometimes referred to as the special ident pulse, occurs when the pilot presses the ident button. A pulse is attached to the coded data string after the *F*2 pulse, and is spaced two spaces away from the *F*2 pulse. This is the special pulse that allows the air traffic controller to specifically identify a given aircraft on his or her radarscope. See Figure 10.13.

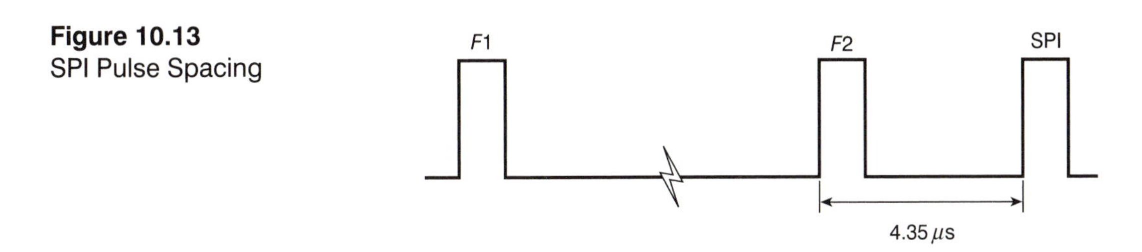

Figure 10.13
SPI Pulse Spacing

MODE *C* OPERATION

Mode *C* altitude operation occurs when the aircraft is interrogated by Mode *C*. The aircraft must be equipped with either an encoding altimeter, or with an altimeter digitizer. The encoding altimeter has a dual function. It displays altitude information to the flight deck and provides the digital conversion for the Mode *C* transmissions. The altimeter digitizer is a remote unit that converts pressure altitude into electronic code for Mode *C* transmission. The altitude data could also be data from a central air data computer (CADC). In either case, when the aircraft has been interrogated by the ground station in Mode *C,* the transponder must recognize the interrogation as valid, and then reply with altitude information. The altitude encoding is electronically accomplished with the use of the Gray code. The **Gray code** is a number system designed to change only one bit at a time. The Binary, or any other number system, can have two, three, or even more bits, changing at once. For example, when the number 79 increments to 80 there are two bit changes. The 7 changes to an 8 and the 9 changes to a 0. The Gray code, in a modified form used in aviation, is referred to as the Gillham code. This unique code is used to transfer the pressure altitude of the aircraft to the ground station. (See Appendix D, Altitude Reporting Code Chart for Transponder Mode *C;* Appendix E, Gray Code and Conversions; and Appendix F, Gray to Binary.)

To quickly recap:

- The Mode *A* reply is in response to a Mode *A* interrogation and transfers the 4-number code set by the pilot.
- The Mode *C* reply is in response to a Mode *C* interrogation.

  In Mode *C,* the interrogator alternately changes between *A* and *C* with the reply pattern being Mode *A,* Mode *C,* Mode *A,* Mode *C,* and so on. This transmits the pilot-set 4-number code and altitude pressure.

BASIC AIRBORNE TRANSPONDER DIAGRAM

The basic airborne transponder shares a common transmitter and receiver antenna. The duplexer acts as an RF switch to allow for transmission or reception. The decoder determines if the ground station has sent a valid interrogation. The decoder determines three things: whether the *P*1 pulse is first, whether the *P*2 pulse is 9 dB lower than the *P*1 pulse, and the mode of the interrogation. If the decoder recognizes the interrogation as valid, then it allows the encoder to function. The encoder gathers altitude data and the code data that the pilot has set on the control head. This altitude data could come from an Air Data Computer, from an altitude digitizer, or from an encoding altimeter. The data are then organized into a reply code. This reply code is modulated, amplified, and transmitted out through the antenna, as illustrated in Figure 10.14.

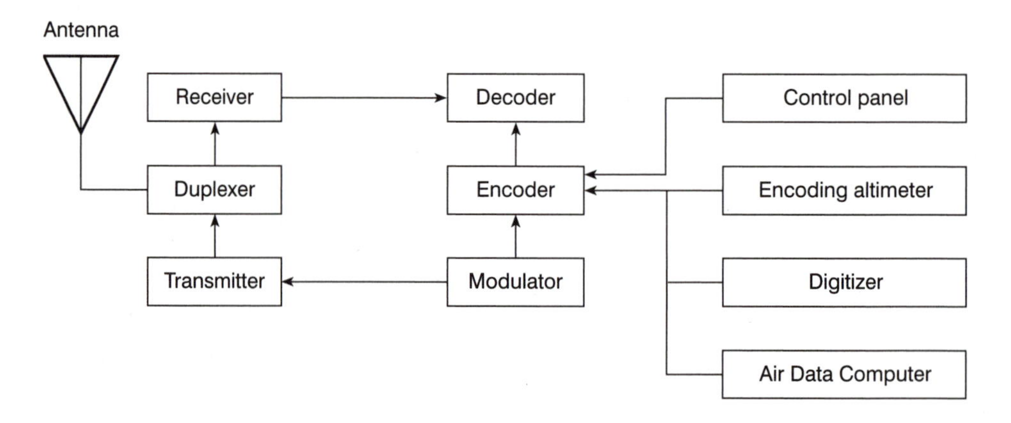

Figure 10.14
Block Diagram of Typical Airborne Transponder

TECHNICAL DETAIL—AIRBORNE TRANSPONDER SYSTEM DIAGRAM

This section contains more electronic details about the system operation. Figure 10.15 shows the block diagram of a typical transponder. The pilot provides the input functions of the mode select, code select, and the ident. The interrogation signal is picked up by the antenna. A single antenna is used for both receive and transmit function, so the **diplexer** (or duplexer) is used to direct the received radio energy to the receiver circuit. The diplexer also prevents high energy transmit power from damaging the sensitive receiver circuitry during transmit times. The received energy is then processed in a superhetrodyne receiver. The 1030 Hz receive signal is mixed (or heterodyned) with the local oscillator signal to produce a difference frequency of 60 MHz. It should be noted there is a 60 MHz difference between the receive frequency of 1030 MHz and the transmit frequency of 1090 MHz, so a local oscillator may be used as a source signal for the transmit function. However, this is seldom done. Usually the transmitter signal is created by a pulsed cavity oscillator.

The receiver intermediate frequency signal is logarithmically amplified using the process of successive detection. The detected output is then fed to

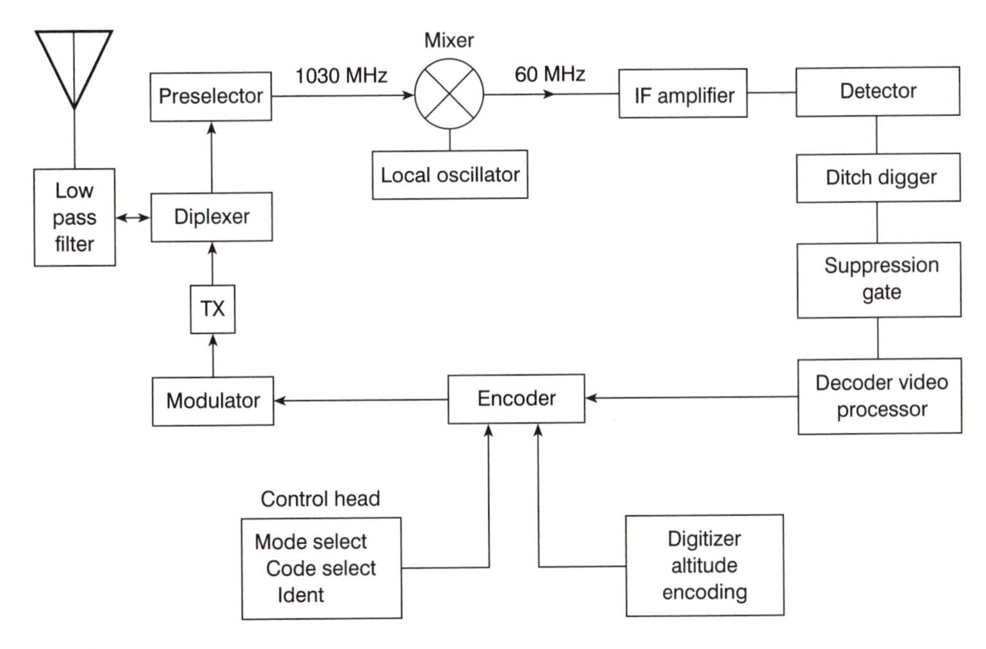

Figure 10.15
Typical Block Diagram of Transponder

the ditch digger circuit. The ditch digger compares the size of the *P*1 pulse to the size of *P*2 pulse, and then makes a decision to either allow or disallow the *P*2 pulse. At the output of the ditch digger circuit, *P*2 will either be on (full amplitude), or will be off.

A suppression gate will not allow decoding of any received pulses during conditions of side lobe interrogations, standby, transponder transmit time, or DME transmit time.

The video processor determines the mode of operation and loads either Mode *A* reply pulses from the control head, or altitude pulses from the digitizer into the encoder register. The video processor also triggers the system into transmit mode. The encoder feeds data to the modulator, which triggers the cavity oscillator transmitter. Transmitter energy is then directed to the antenna via the diplexer. The cavity oscillator signal is typically rich in harmonics (side frequencies), so a low pass filter is used to prevent interference with other aircraft systems.

MAJOR DISADVANTAGES OF ATCRBS

ATCRBS has experienced some difficulty with an over loaded system, synchronous garble, and fruiting. The ATC must establish voice communication with each and every aircraft. Not only does this communication happen when aircraft enter control zones but communication could happen as many as five times while a single aircraft passes through a control zone or completes a landing. ATCTX with altitude Mode *A*/*C* alleviates some of this voice communication.

MODE *S* TRANSPONDER

As of 1992 there is a legal requirement for all new aircraft to be equipped with a Mode *S* transponder. Mode *S* stands for "select." The development of the Mode *S* has taken some time, but it is now a reality. The major reasons for the development of Mode *S* are to eliminate some of the ATCRBS problems such as fruiting, synchronous garble, and a limited number of available codes (4096). Mode *S* allows aircraft identification without voice communications, and provides a ground/air data link capability. This data link allows an individual aircraft to respond to a discrete interrogation and gather data such as weather services, airport services (ATIS), and runway surface winds. It reduces the ground controller voice communication time. No longer will the ground controller need voice communication for altitude, heading, type of aircraft, etc. With Mode *S,* each aircraft will be identified with its own code. This code will be on a PROM chip, which can be attached permanently to the aircraft, perhaps on the back of the transponder mounting tray. In the event of a transponder change, the transponder interfaces with this chip, and the aircraft is still identified by the unique algorithm containing the aircraft's registration number.

BRIEF SYSTEM OPERATIONAL OVERVIEW

Here's how it works. An aircraft enters the ATC **Control Zone.** The ATC interrogates the aircraft using Modes *A, C, S* on a cyclic basis. If the aircraft is equipped Mode *A,* its transponder replies Mode *A.* If the aircraft is equipped Mode *A*/*C* altitude, the transponder replies Mode *A* and *C.* If the aircraft is equipped Mode *S,* the transponder responds Mode *S.*

In the Mode *S* interrogation the first contact with the "new" aircraft to the zone is the "All Call" function. Any aircraft "listening" will respond. Once the aircraft has been identified through the All Call interrogation and the subsequent transponder's reply of its algorithm, the ATC then knows the aircraft's registration, and is able to predict this aircraft's position. Once a Mode *S* transponder replies to an All Call interrogation, the ground station is aware of the specific aircraft's identity. This is because each aircraft has a unique address that is part of the Mode *S* reply. The next interrogation of that aircraft will be a very specific Mode *S* uplink code. This code contains a coded command to lock out any further All Call replies. This keeps the amount of radio traffic to a minimum, and the ATC is now able to interrogate a specific aircraft at leisure, while still knowing each aircraft's registration, heading, speed, etc. Now the aircraft is interrogated with an uplink code format, and specific data are now transferred. There are 25 different uplink code formats that can be used. This is akin to the United Nations, where 25 different languages are being used by 25 different people—at the same time—all through one interpreter, in our case the ATC computer. At the present time only 4 uplink codes are in use.

QUESTIONS

1. What is the SSR interrogation frequency?
2. What is the transponder reply frequency?
3. What is a side lobe interrogation?

4. What is a main beam interrogation?
5. How does the transponder distinguish between the main beam and side lobe interrogations?
6. Altitude information is supplied through what operational mode?
7. What information is transferred through the transponder?
8. What is the ATC transponder audio output?
9. Using the altitude Reporting Code Chart (in Appendix D), sketch the transponder code sequence for 26,500 ft of altitude.
10. Why is the Gray code used for altitude encoding?
11. The word RADAR is an acronym. What does it stand for?
12. What is the function of the SPI pulse in a transponder reply?
13. When in Mode *C*, what information is transmitted to the ATC ground station? What is the transmit sequence for these parameters?
14. What does squawk ident mean? Who says it? What happens?
15. What is the purpose of pulses $F1$ and $F2$?
16. What is the optimum location of the ATCTX airborne antenna?
17. How many reply codes are available to a Mode *A* interrogation?
18. How is skin return measured (units)?
19. Why does the transponder require a suppression bus?
20. Why does ATC exist?
21. What is a Control Zone? Approximate size?
22. How does the pilot know if the transponder is working?
23. What is an ATC clearance?
24. Sketch and label the transponder pulse code for an onboard emergency (7700).
25. Show the pulse code for a selected code of 2563.
26. What does the *S* stand for in Mode *S*?

CHAPTER 11

Traffic Alert and Collision Avoidance

TCAS

Introduction

This chapter deals with the underlying concerns of Air Traffic Control and the evolution of TCAS as a result of those concerns.

Objectives

State current status of Air Traffic Control

State why there is a need for TCAS

State the purpose of the system

State what the acronym TCAS means

Identify the levels of TCAS

Identify the priority levels

State the approximate size of the "monitored airspace"

State the audio and visual outputs

Identify TA and RA

Speculate on how ATC facilities will change within the next 10 years

Basic Concept

TCAS is the acronym for Traffic Alert and Collision Avoidance System and is an avoidance system that alerts the crew if any aircraft should enter into the aircraft's envelope of monitored airspace.

THEORY OF OPERATION

Traffic Alert and Collision Avoidance System (TCAS) supports the concepts of "see and be seen." On a priority basis, the pilot probably fears onboard fire the most, and midair collision next. TCAS is designed to function as an electronic aid to human factors engineering techniques. In the past these have included such items as removing the center post on windshields and making larger windows to provide for increased visibility. The original TCAS was introduced in 1970, but had been delayed by the FAA. As of 1994 all commercial air carriers must be equipped with TCAS. TCAS is a system that provides the flight crew with traffic alert information and in its more enhanced operation also provides the crew with recommended course of action. TCAS does not override ATC requests but acts as an aid to the pilot in safe routing of his aircraft.

TCAS is a secondary radar facility. The airborne TCAS system interacts with the transponder system (Mode *S* and ATCRBS). TCAS is virtually an ATC that is airborne. The TCAS system does not directly tie in with the ATCRBS. The system:

- is independent but compatible with existing ATC facilities
- interrogates other aircraft transponders
- replies to other interrogations
- displays to the flight crew all transponder-equipped traffic in the vicinity
- provides the flight crew with audio alerts

TCAS is the incorporation of a number of systems and devices coupled together to function as a surveillance system and as an avoidance system. TCAS I is a proximity warning only—it assists the pilot in visual acquisition of an intruder and is used in general and commuter aviation. TCAS I will actively interrogate nearby aircraft transponders and provide bearing and range information as an advisory. The system tracks aircraft and computes convergence time and altitude. It uses the solution to these calculations to classify the level of threat of the target aircraft. TCAS I will not provide climb or dive commands and the flight crew must determine the evasive action required.

TCAS II provides Traffic Advisory and Resolution Advisory, making recommendations to the flight crew in the vertical mode only. It is used by airlines and business aviation.

TCAS III is under development but will eventually provide the flight crew with Traffic Advisories and Resolution Advisories in the vertical and horizontal modes. **Traffic Advisory (TA)** is an audiovisual alarm to the flight crew. The Traffic Advisory is actuated when TCAS determines that there is air traffic on a possible collision course. The **Resolution Advisory (RA)** provides the flight crew with an electronic solution to the traffic problem.

Naturally the TCAS provides no protection from any aircraft with no transponder. TCAS III will provide horizontal avoidance commands.

The TCAS displays traffic position and altitude. The electronics create a "window" or an envelope of protected airspace that is about 3 mi in diameter and about 750 ft deep. The interrogation range is about 45 NM ahead and 10 NM behind.

The computer categorizes the targets on a priority basis. Priorities are:

- no threat
- proximate traffic
- potential threat
- immediate threat

At this time an aircraft is under ATC control from the time it taxis from the hangar until it arrives at its destination. The actual enroute control is with **Terminal Radar Approach and Control—TRACON** (US), and with **Joint Enroute Terminal System—JETS** (Canada).

The current status of the Air Traffic Control System is antiquated equipment and overloaded operations. The FAA in the United States and Transport Canada are both in serious trouble, with air traffic predicted to double in the next 10 years. The air traffic controller is subjected to so much information that he can react to problems only as they arise. This procedure has resulted in lost time per flight of between 3 and 15 min. In 1991 an estimated 75, 000 hours were lost due to air traffic control procedures.

The risk of aircraft collisions increases dramatically when bad weather, crowded airports, antiquated equipment, and human factors are considered.

The solution is to move the industry into automated computer control systems that will reduce delays and minimize collisions. The FAA is currently working on a system called **AERA—Automated Enroute Air Traffic** control, which is a $32 billion computer program. This program will solve the collision problem with predicted 99.99% accuracy. AERA will evaluate the aircraft's location, altitude, and speed and add the wind factor. It will then make a recommendation based on predicted outcomes 20 NM ahead of occurrence. It is also being designed to solve traffic flow problems, particularly with converging traffic to the final approach.

The computer solution takes place in two stages. The first stage is automated databases for tracking flight information and automated radar. The second is intelligent programs that are estimated to perform 80% of the ATC decision. The major advantage here is the ability of the computer to predict. It will identify and resolve conflicts 20 min prior to their eventuality. One intelligent program, CTAS (Center TRACON Automatic System), will predict optimal spacing between aircraft in preparation for approach. Other intelligent systems will be voice recognition and direct links between the ground and aircraft computers with no human intervention. Such things as routine landing clearance and updated weather will solve many of the overloaded voice VHF communication problems. All of this aside, the computer can "fly" the aircraft more precisely than can the human.

TCAS SYSTEM COMPONENTS

Figure 11.1 summarizes the TCAS system components, which are outlined below in more detail:

- Mode *S* Control Panel (selects and controls all TCAS elements):
  - TCAS computer
  - Mode *S* transponder
  - TCAS displays

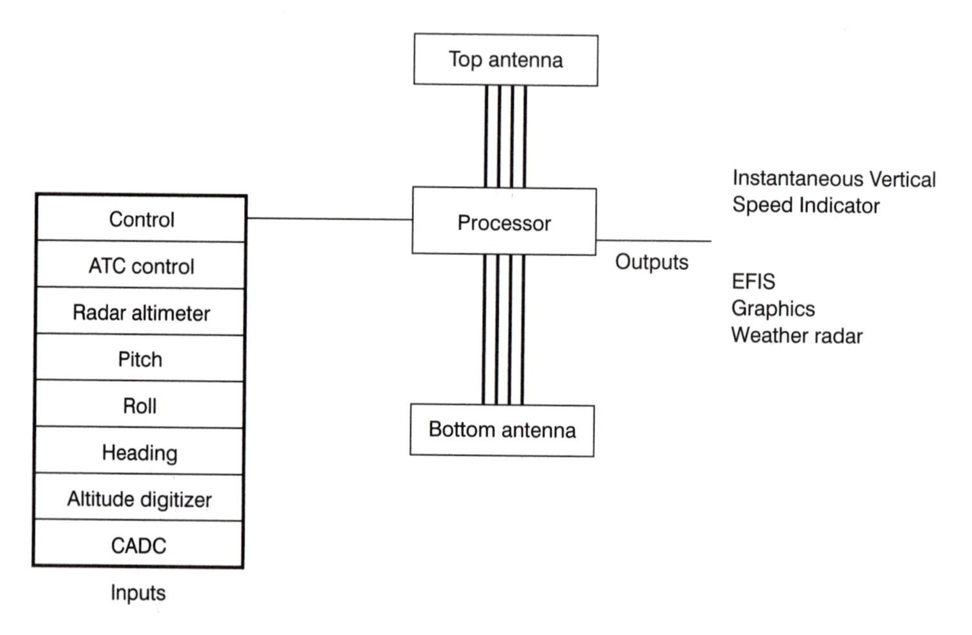

Figure 11.1
TCAS System Components

- Mode *S* Transponder:

  performs *A*, *C*, and *S* transponder function

  TCAS to TCAS communication for complementary resolution advisories

- TCAS computer unit:

  performs monitoring airspace surveillance

  intruder tracking

  threat detection

  advisory generation

- Traffic Advisory Display (TA) (TCAS I):

  depicts traffic relative position

  display can be dedicated or shared with weather radar possibly on EFIS or flat screen

- Resolution Advisory Display (RA) (TCAS II and TCAS III):

  RA display is standard Vertical Speed Indicator (VSI) modified to indicate the vertical rate that must be achieved to maintain safe separation. The RA display has red and green indicator lights. In order to comply with the RA, flight crew must keep VSI needle out of the red zone.

- Aural:

  TA and RA are supplemented with synthetic voice advisories

  A target infringing on the protected airspace envelope will turn yellow and aural Traffic Advisory (TA) "TRAFFIC, TRAFFIC, TRAFFIC" will sound.

  A target as an immediate threat—the computer has computed that evasive action is required (called a Resolution Advisory (RA))

  The visual indication will turn red and aural "CLIMB, CLIMB, CLIMB" will sound.

TA DISPLAY AND ADVISORY

Remember that the function of a TA is to aid the flight crew by detecting intruding aircraft. After the TCAS system detects the intruding aircraft, the system helps the pilot to "visually acquire" or "locate in space" (see) the intruder. The data in the system are updated once every second and may include relative altitude information or actual altitude. The display shows:

- TA display
  own aircraft
  shown as an airplane or arrow
  white
- Nonintruding traffic
  open diamond
  white
  data tag (available on altitude reporting traffic)
- "Proximate" target
  with 6 NM and 1200 ft vertical
  white diamond—filled
  data tag
- Potential threat
  yellow solid circle
- TA
  red filled square
  data flag
  TA shows conflicting traffic is approximately 40 sec away
  Traffic threat has traffic 25 sec away.

A few of the current suppliers of TCAS are Allied Signal Aerospace, BF Goodrich Flight Systems, and Trimble Navigation. These systems are all operational on ARINC 429 data bus protocol.

IMPLEMENTATION

In the United States, TCAS II was implemented, installed, and functional by 1993.

Current training for airline pilots has mushroomed, using computer-based training in order to update the 60,000 pilots affected. As of February 1993, 70% of US airlines and 650 corporate aircraft were equipped with TCAS. In 1994, the 10–30 seat aircraft were required to install TCAS I.

ADVANTAGES OF TCAS

TCAS has several advantages:

- oceanic track separation
- increased airport capacity
- aid ATC in speed control and separation
- legally allows for increased air traffic and closer spacing out of ATC

PRECAUTIONARY NOTES

When testing transponders on the ground, use caution, because such testing appears real to another airborne interrogation. This may cause aircraft just landing or taking off to make rapid and dangerous avoidance maneuvers.

QUESTIONS

1. TCAS stands for Traffic Alert and Collision Avoidance System. (True or False?)
2. Interrogation range of the TCAS system is about 45 NM. (True or False?)
3. Targets are categorized on a priority basis. (True or False?)
4. Two warning methods are initiated from TCAS to the flight crew. (True or False?)
5. There is absolutely no problem in testing TCAS and transponder when on the ground. (True or False?)
6. The envelope of "protected" airspace around the aircraft is about 3 mi in diameter and 750 ft deep. (True or False?)
7. What is the meaning of the acronyms:

   JETS?

   TRACON?
8. State three reasons for updating current ATC facilities.
9. Explain current status of Air Traffic Control in the USA and in Canada.
10. Explain the 10-year plan for ATC.

CHAPTER 12

Long Range Navigation

Loran C

Introduction

This chapter deals with the basic principles of the Loran C navigation system.

Objectives

Identify the basic principle of navigation with Loran C

Identify frequency of operation

Explain how the unit functions

Identify the block diagram of the units

Explain what the pilot sees and uses with the Loran C

State the significance of the numerical designation of a Loran chain

Explain transmit sequence from a Loran chain

Describe the transmit format for master and slave stations

Basic Concept

Loran C is a low-frequency long-range radio navigation aid.

THEORY OF OPERATION

The original Loran was developed during World War II to provide for an extensive worldwide system for navigation over water. **Loran** is an acronym for *Lo*ng *Ra*nge *N*avigation. It is a long-range low-frequency hyperbolic radio navigation system that uses time intervals instead of exact distances. Originally designed for military use, it has been modified and is now used extensively by almost everyone. Its current version is called Loran C.

The Loran C system is a low-frequency system that is effective at all times over extremely long distances. The Loran C navigation system can provide for an accuracy of less than 1 mi over a distance of 1500 mi. The operation of the system is such that the elapsed time between the reception of a master signal and reception of signals from two or more slave stations provides two or more hyperbolic curves, and the intersection of these curves is the location of the aircraft.

Through experimentation it has been found that the frequency of 100 kHz provides the best propagation characteristics over both land and sea when very large distance is involved. This low frequency allows curvature of the earth signal. Loran stations typically transmit at power levels from 400 kW to greater than 1000 kW.

The complete Loran system is made up of 22 *chains.* Each chain consists of a master transmitter and two to five *slave* or secondary transmitters. Each chain is identified by the use of a four-digit number that is representative of its **Group Repetition Interval (GRI).**

The aircraft receiver uses the time difference between the reception of the master signal and the secondary signals to develop lines of position (LOP). The location of the aircraft can then be determined. The minimum navigation requirement is three stations, a master and two secondaries. These three stations are called a *triad.* Three stations are required to determine position.

The receiver then displays latitude and longitude in degrees, minutes, and hundredths of minutes. This provides precision of ± 100 ft on the baseline.

The Loran C receiver rejects sky wave contamination due to ionospheric and atmospheric variables. The sky wave transmission is not accurate for Loran. Loran C uses ground wave propagation.

The transmitters are fixed land-based units organized in groups called chains. Each chain has a master and two to four secondaries, which are designated as W, X, Y, Z. The chain is designated by its GRI. Each chain has a unique GRI. The GRI is defined as the time between subsequent master station group transmissions. The GRI is measured in μs. For example, the US West Coast GRI is 99,400 μs. Therefore the chain designation is 9940. Consider the chain that is called the Canadian West Coast GRI 5990, which consists of M, X, Y, Z stations with the master at Williams Lake, British Columbia; X at Shoal Cove, Alaska; Y at George, Washington; and Z at Port Hardy, British Columbia.

Remember that the 5990 chain repeats the master pulse group every 59.9 milliseconds (ms) or every 59,900 microseconds (μs). The master station transmitter transmits a series of pulses and then each secondary transmitter transmits a group of pulses after a predetermined time delay. Figure 12.1 shows the system transmission sequence, with the arrows indicating the predetermined time delays and the long topmost line indicating time.

These pulses are actually 27 sinusoidal waves in an increasing amplitude to about the eighth wave, and then decreasing, as shown in Figure 12.2.

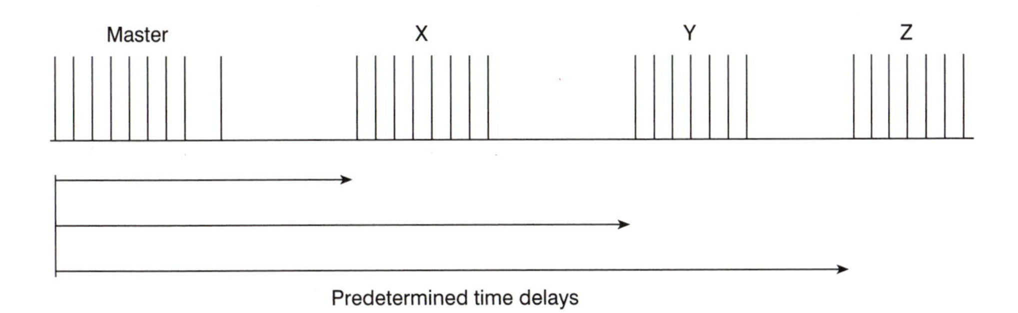

Figure 12.1
Loran System Transmission Sequence

Figure 12.2
Pulse Formats

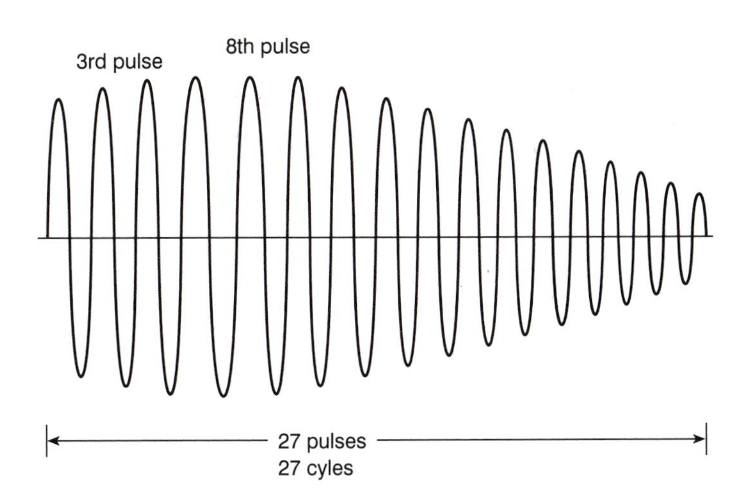

Each sinusoidal wave has a 27 μs period. The third wave is the one that is actually used for timing purposes in the normal mode because the third pulse occurs within the first 30 μs of the pulse. Anything after this time is subject to distortion from reflections, as noted in Figure 12.3. The eighth pulse is the largest and therefore has the best signal-to-noise ratio, but is susceptible to distortion due to reflections from the atmosphere. For this reason, the eighth pulse is used for extended range mode with the understanding that reduced accuracy results.

In Figure 12.4, the vertical pulse group lines are actually the same as the vertical lines in Figure 12.1; this figure shows greater detail.

Figure 12.3
Wave Propagation Characteristics

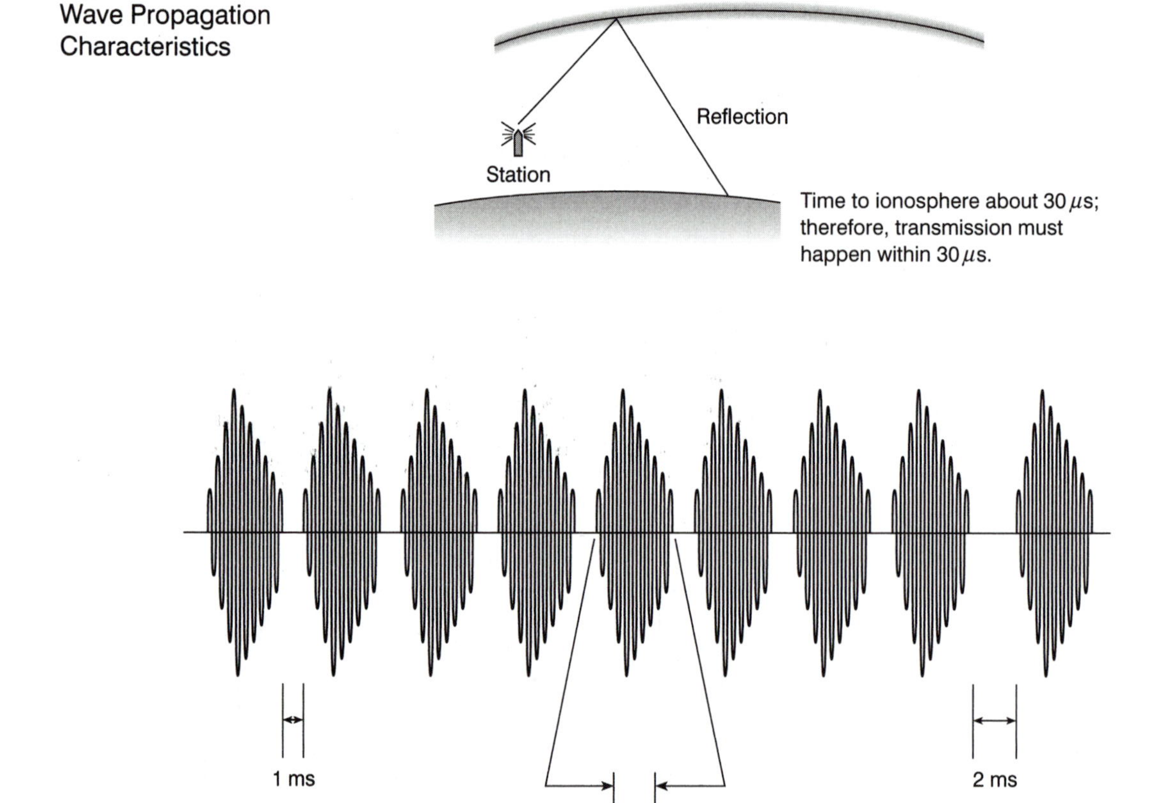

Figure 12.4
Master Pulse Group Format

LORAN RECEIVER

The Loran receiver examines

- stored information regarding GRI
- stored information regarding secondaries
- leading edge of pulse train

The Loran receiver is a position-determining device where all data are calculated by continuous updating. The Loran receiver does a sampling from a chain, then selects the triad with the best signal. The latest and greatest receivers are capable of selecting from more than one chain at a time. Receivers of this nature are capable of tracking 12 stations at a time and some can have 40,000 elements in the memory. They can be programmed to hold 10 flight plans and up to 20 Waypoints. The microprocessor also allows storage of airports by name, city, or country if the pilot does not know the station identifier.

The Loran antenna can take a wide variety of shapes and formats. They can be E or H polarized, can be long or short whip types, or can share the ADF antenna with a signal coupler, although this procedure is not recommended.

NAVIGATION (A FEW REMINDERS)

Latitude measures distance in degrees north and south from the equator, which is at 0°. Parallels of latitudes are such that each minute always equals 1 NM, as shown in Figure 12.5. Longitude indicates the east-west meridians, where each minute of longitude is equal to 1 NM at the equator but is reduced in length toward the poles. This is why the vertical on the map must always be used for caliper settings.

Figure 12.5
Longitude Characteristic

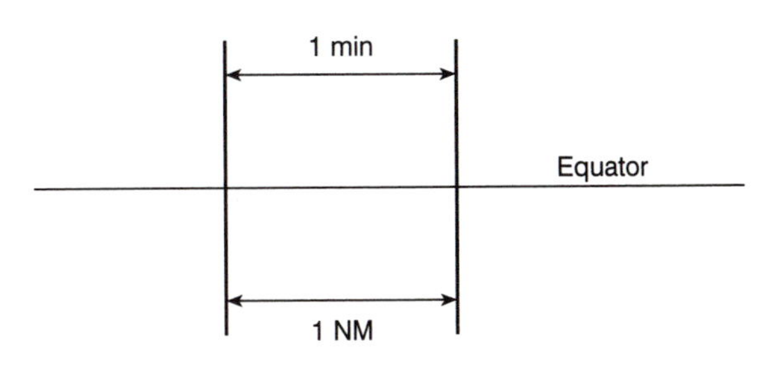

ACCURACY

Absolute accuracy is the ability of the Loran C unit to select and determine true geographical position. Predicted and actual error are ±100 feet on the baseline.

When the aircraft is located in a straight line beyond a secondary from the master, the line of position is referred to as *baseline extension.* Baseline extensions are not usable because the lines of position are nearly parallel. In the example in Figure 12.6, the signal from station W must be deselected or accuracy will be poor.

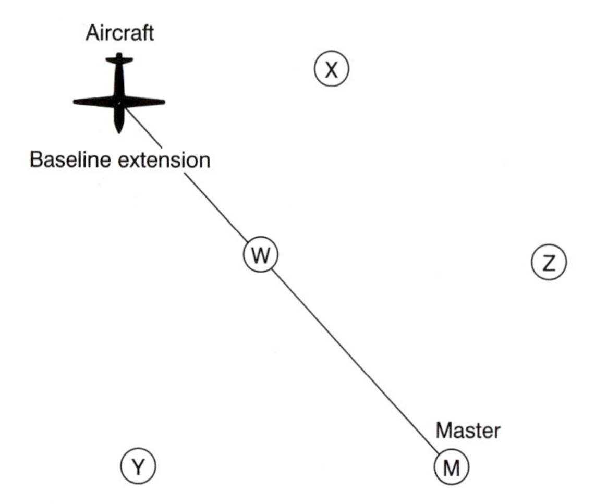

Figure 12.6
Loran Chain

CURRENT STATUS

The release date for charts for the nonprecision approaches using Loran C was 1990. There are no approved IFR approach receivers available. The IFR Loran unit must be certified for approach use. All of the Loran C receivers that have been installed to date are for VFR use and about 10% are for enroute and terminal use. None have been approved for approach usage. This situation exists primarily because Loran C units have a limited capability to inform the flight crew of malfunction in less than 10 sec, which is the minimum requirement for precision approach malfunction detection. Typical Loran C units have about 60 sec warning failure times.

Loran C is *not* approved as a sole use navigation system.

Transmission frequency of 100 kHz is subject to static and as a consequence, the aircraft installation must be done by approved personnel who will ensure that all static problems have been reduced and electrostatic skin mapping has been done to further reduce the effects of static problems. Skin mapping is a surface analysis of the aircraft's magnetic fields to find where antennas interfere with each other, or where there are unusually large areas of magnetic influence that interfere with normal transmit/receiver operation. As Loran is an area navigation system, the loss of transmitter is critical. If Loran C were approved as a sole use navigation system and many aircraft were simultaneously using the chain, then loss of the chain would mean many lost aircraft.

WAYPOINTS

Loran units have Waypoint coordinates for many navigation aids such as VORs, airports, etc. It is possible for the pilot to enter a real or fictitious Waypoint. During Loran C navigation there is constant navigation information given to the flight crew such as bearing, range, speed, and time.

HYPERBOLIC LINE OF POSITION

The navigation form is hyperbolic due to the time delay interval between master pulse and X station pulse. If your position is such that you receive a master signal, and then receive another signal from X Ray station 12,725 μs later, there are an infinite number of possible positions where you could be. The number of possible positions exist on an imaginary hyperbolic curve. We say that we are located somewhere on this line of position (LOP).

To resolve the ambiguity of exact position on the hyperbolic curve, a third station and its timing are required. As in all navigation solutions, best accuracy is achieved when the LOPs cross at nearly right angles.

The master signal can be compared to several slave signals because there is a fixed time delay unique to each chain between the transmission of the master and the transmission of the slave. This time delay is stored in the memory of the receiver. The time delay plus a reasonable propagation delay will form a window of possibility in time within which the reception of the slave signal must fall.

QUESTIONS

1. What is the carrier frequency of Loran C?
2. The Canadian West Coast Chain is designated as 5990. What does the 5990 denote?
3. Why is there a delay between the master and slave transmissions in a Loran C chain?
4. Loran is an acronym. What does it stand for?
5. What is a triad?
6. How many chains exist in the world?
7. How many masters and how many slaves are required for accurate navigation?
8. Can Loran be used as a substitute for ILS approach? Discuss.
9. What does GRI stand for?
10. How does the receiver display the aircraft position?
11. What is the accuracy of a Loran receiver?
12. Loran uses what kind of wave propagation?
13. How are secondary transmitters designated?
14. Explain skin mapping. Why is this necessary?
15. How is one Loran chain distinguished from another?

CHAPTER 13

Omega Navigation System (ONS)

Omega

Introduction

This chapter deals with the basic principles of the Omega Navigation System.

Objectives

Identify the theory of operation of the Omega system

State the frequencies of operation

State the number of stations in the system

State how many common and how many unique frequencies each station has

State the total time duration for a complete transmission

Describe Lane

State modes of operation

Basic Concept

Omega is a worldwide, low-frequency, long-range, hyperbolic navigation system.

OMEGA GROUND STATIONS

The **Omega Navigation System (ONS),** or simply **Omega,** is a hyperbolic navigation system that has eight ground stations. Ground stations provide bursts of continuous wave energy in the VLF frequency band. These signals fall within a timed format. The system is operating on a time share basis with stations transmitting phase synchronized signals. All stations use stable cesium or rubidium atomic clocks in order to maintain precise time measurements. Each station is identified with its own signature frequency. These ground stations provide global coverage because the nature of the low frequencies utilize the earth's atmosphere as a waveguide. The transmitters also have high output power (around 10 kW). The eight stations in the Omega system are listed in Figure 13.1. Appendix G lists VLF stations available for Omega supplemental usage.

Each station transmits four common frequencies, and one frequency that is unique to that station. The unique frequency is called the signature frequency. The eight stations continuously change frequencies in a unique pattern. This transmission process provides for station identification.

The four common frequencies are:

10.2 kHz

13.6 kHz

Letter	Number	Location	Country responsible for maintenance
A	1	Aldra	Norway
B	2	Monrovia	Liberia
C	3	Haiko, Hawaii	United States
D	4	La Moure, North Dakota	United States
E	5	Réunion*	France
F	6	Golfo Nuevo	Argentina
G	7	Woodside	Australia
H	8	Tsushima	Japan

*Réunion, an island in the Indian Ocean, is an overseas department of France.

Figure 13.1
Omega Stations

11.33 kHz
11.05 kHz

The signature frequencies and locations of the stations are given in Figure 13.2.

Figure 13.3 shows the transmission format for the whole system.

The overall signal duration is 10 sec. The transmission format is of unequal segments of time duration and at different frequencies. Note that the pattern of transmission for each station is such that the signature is transmitted twice each 10-sec cycle, and that the duration of signature is different for each transmission. This arrangement allows the airborne Omega computer to synchronize with the Omega format and identify specific ground stations.

The major disadvantages of Omega are that navigation depends entirely on ground-based stations, and the ground-based stations are very expensive. Omega is also very susceptible to interference from precipitation static.

There is a group of VLF stations that can also be used as a supplement to the Omega chain. These VLF stations are synchronized to provide a time reference for shipping, and can be used as an alternative input signal to Omega. These VLF stations are actually for submarine communications and their ability to be used for global navigation is a lucky coincidence.

Letter	Number	Location	Unique or Signature Frequency
A	1	Aldra, Norway	12.1 kHz
B	2	Monrovia, Liberia	12.0 kHz
C	3	Haiko, Hawaii, U.S.A.	11.8 kHz
D	4	La Moure, North Dakota, U.S.A.	13.1 kHz
E	5	Réunion*	12.3 kHz
F	6	Golfo Nuevo, Argentina	12.9 kHz
G	7	Woodside, Australia	13.0 kHz
H	8	Tsushima, Japan	12.8 kHz

*Réunion, an island in the Indian Ocean, is an overseas department of France.

Figure 13.2
Omega Stations and Frequencies

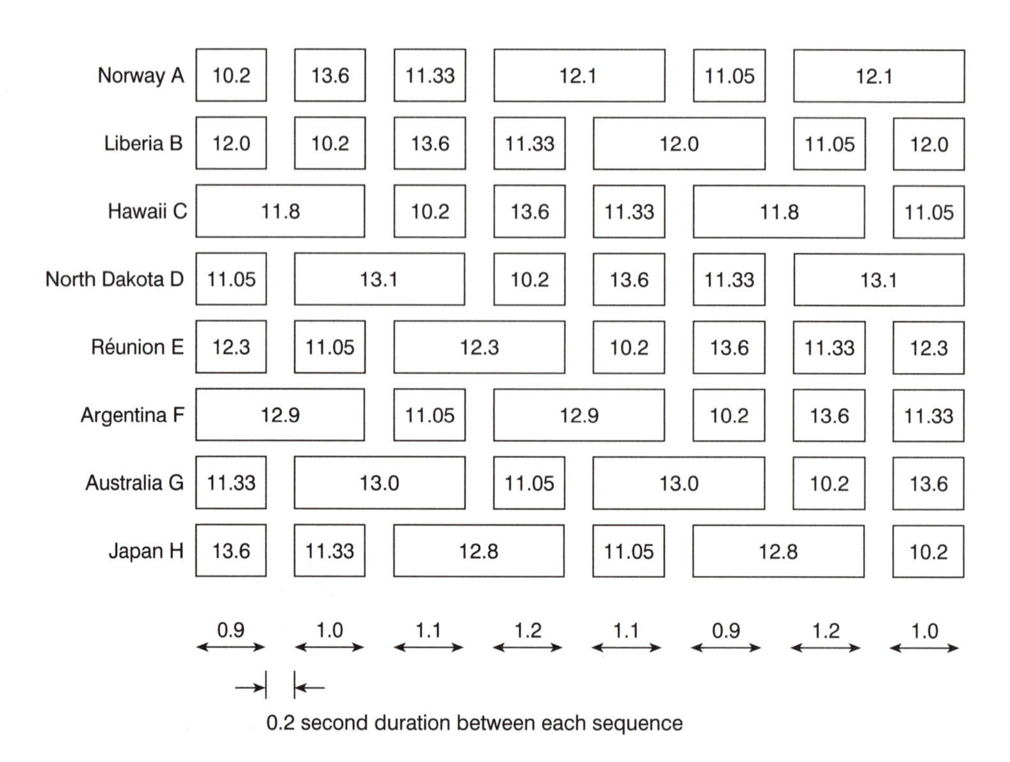

Figure 13.3
Omega System Overall Transmission Format

THEORY OF OPERATION

The Very Low Frequency radio wave is transmitted in every direction, like the waves created in a pond from a thrown rock. As with all radio waves, there is a constant relationship between the frequency and the wavelength. This has been defined as:

$$\lambda = \frac{300 \times 10^6 \text{ m/s}}{\text{frequency}}$$

where lambda (λ) is the wavelength of the sine wave and
300×10^6 m/s is the speed of light

For example, a frequency of 3 MHz has a wavelength of 3 mm. Consider the signature frequency of Hawaii at 11.8 kHz. It has a wavelength of 13.73 NM! The wave repeats itself every 13.73 NM. Omega creates **Lanes** by calculating a standing wave pattern from two signals that come from different points on the earth. The Lanes relate to the frequency or wavelength of the signals, or the difference between two signals of different frequencies.

The baseline is determined as the **Great Circle** distance between two ground stations. The Omega system compares the received signal with a very accurate onboard time reference in order to derive a phase measurement. Although two stations never broadcast the same frequency at the same time, the Omega system can compare the same frequency from two stations as if they occurred at the same time, thus creating the equivalent of a standing wave. The mathematical relationship between the frequency of a waveform and its standing wave (as if it were reflected) is a 1:2 ratio. That is, the standing wave repeats itself every 180° of the RF wave.

Frequency	Wavelength	Lane width
10.2 kHz	15.88 NM	7.94 NM
11.05 kHz	14.66 NM	7.33 NM
11.33 kHz	14.29 NM	7.14 NM
13.6 kHz	11.91 NM	5.95 NM

Consider two stations. Each is transmitting according to the system pattern shown previously. As a consequence there is an infinite number of concentric circles of energy radiating out across the Pacific. This energy is of different frequencies and different wavelengths depending upon the segment of the Omega format being transmitted at that time. The Omega unit has one independent receiver for each frequency and compares the phase of these signals to its onboard reference. It then derives a number representing the position within the Lane for each frequency. These numbers are entered into the computer's memory. As the aircraft moves, the phase relationship changes and the Omega system keeps track of the changing numbers. The software program of the navigation system uses the numbers in the memory with one of several different mathematical procedures to solve for position.

It must be noted that the Omega cannot tell one Lane from another, but only the changes within a Lane or changes from one Lane to another. This means that the Omega system must have the initial aircraft position entered into the computer at the point of origin. If this information is lost the Omega will become lost. If the received signal is lost due to noise interference (thunderstorm, etc.), the computer navigates from inputs representing speed and distance provided by the air data computer and compass. This is dead reckoning navigation and not as accurate as navigating by the use of received signals. Once the signal is regained, the Omega may be able to interpolate its position once again. Dead reckoning is also used to smooth out the time periods between position updates as the computer is calculating. Update time is usually once every 10 sec.

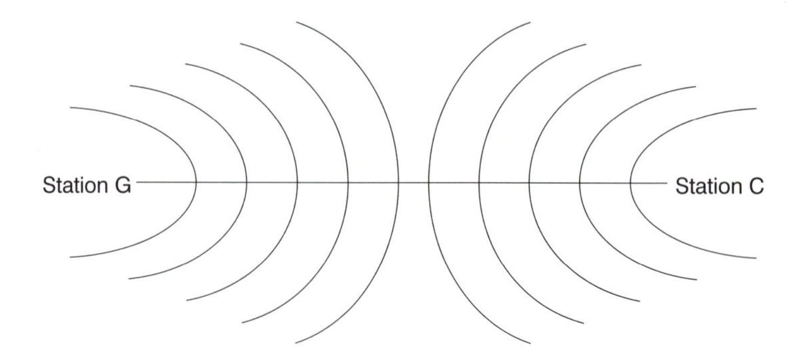

Figure 13.4
Omega Lines of Position

Consider the example of an aircraft flying across the Pacific Ocean. At any given position the aircraft would receive signals, one from Australia (G) and one from Hawaii (C). The result would be a set of Lanes resembling hyperbolic curves symmetrical about a Great Circle line connecting the two stations (baseline), as shown in Figure 13.4.

There is a set of Lanes for each common frequency (not signature frequency) and for each combination of common frequencies for these two stations. This pattern of Lanes is duplicated for each baseline connecting each Omega station. This results in a very complicated crosshatch of Lanes crossing each other at different angles. Aircraft navigation is performed by keeping track of the distance parallel to each baseline that the aircraft travels, and performing a vector addition of these distances to arrive at a bearing and distance change from the last update.

The Omega system will automatically deselect any station greater than a certain distance away because the signal may propagate around the world from either direction. It also deselects a signal from a station that is too close because the signal may arrive distorted by adding the bounced signal with the direct signal.

The computer needs information from at least three stations in order to triangulate and establish position when in the hyperbolic mode. These three ground signals need to have good geometry and need to be quality signals.

OMEGA SYSTEM COMPONENTS

The Omega onboard system contains:

- control display unit (CDU)
- receiver processor unit (RPU)
- antenna and coupler unit (ACU)

The control display unit is the control head located in the cockpit. The pilot can access information and input information to the Omega processor through the CDU keyboard and display unit.

The receiver processor is a computer capable of doing complex navigation calculations. It includes a storage capability—memory. The general functions of the processor are: receive, amplify, compute, interface, display to CDU, and provide autopilot coupling. In addition there is a requirement for accurate time measurement and storage of ground station data, such as position and frequencies.

The receiver determines which stations are within usable range and monitors the quality of the signals received. The processor calculates LOP geometry, and the quality of this geometry, as well as continuously monitoring the number of Lane changes. In general, the processor performs continuous calculation on the overall quality of navigation and must keep track of the aircraft's position. It provides the flight crew with 10-sec updates on the aircraft's most probable position. In the event of receiver fail test, or signal loss test, the receiver is deselected, and the flight crew alerted.

The antenna may be an electrostatic E field antenna (like an ADF sense antenna) or, more commonly, an electromagnetic H field antenna tuned to the VLF band.

The system block diagram in Figure 13.5 shows the typical interconnections in an Omega system.

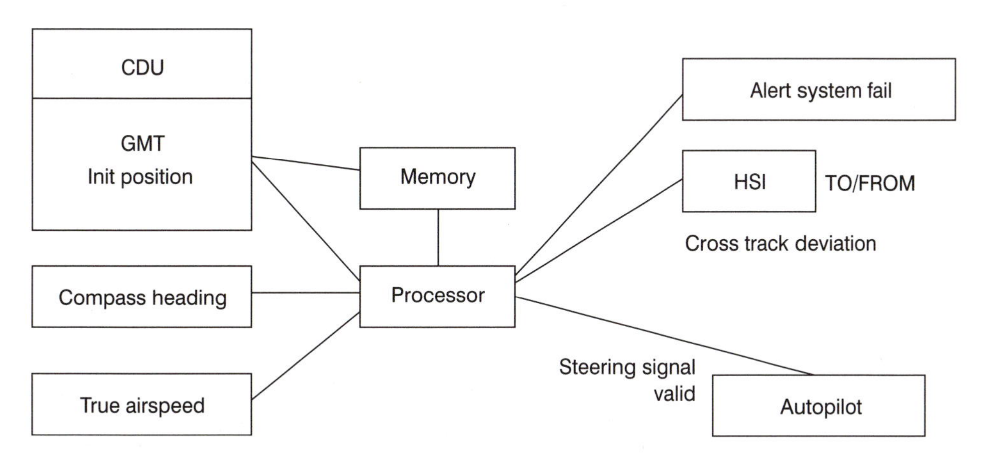

Figure 13.5
Omega System Typical Interconnect

MODES OF OPERATION

There are five modes of operation in an Omega system:

initialize: The initialization process is completed with the aircraft on the ground. The flight crew must enter the actual aircraft position and accurate **Greenwich Mean Time (GMT).** GMT is required because of the diurnal change in propagation. Other data may be entered, such as Waypoints.

hyperbolic: Four or more Omega stations being received. This is sometimes referred to as the absolute mode, and it uses only the eight dedicated Omega ground stations.

range, range, range: Three stations being received.

range, range: Two stations being received.

dead reckoning: Calculations based on (1) initial position, (2) direction of flight, (3) airspeed, and (4) time.

CONDITIONS AFFECTING OMEGA TRANSMISSION

Manufacturers of Omega claim an overall accuracy of ±300 m anywhere in the world. It should be noted that in the VLF band a number of characteristics that will affect the general accuracy of Omega. VLF radio transmission is relatively stable and has predictable propagation characteristics in spite of the many conditions that would attempt to degrade the transmission. Some of these are:

- earth's magnetic effect
- signal attenuation occurring at the polar ice caps
- solar activity
- short and long path distortion
- ionosphere
- season

A change of altitude of the ionosphere is called the **diurnal effect** (see Figure 13.6). When signals pass from day to night, or vice versa, the propagation velocity changes, resulting in a phase shift that must be compensated for. The wave travels through the atmosphere in what could be considered to be a giant wave guide.

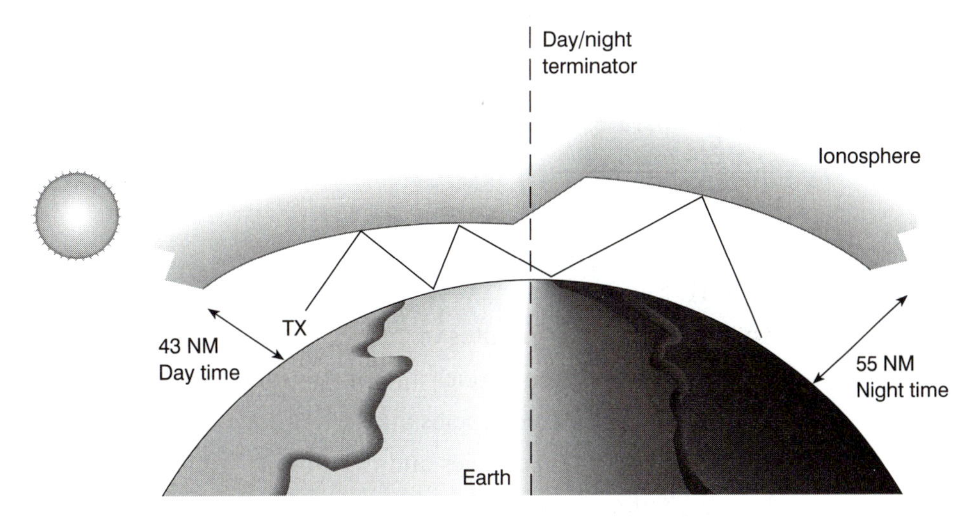

Figure 13.6
Diurnal Effect

If the receiver is within a certain distance of the ground station, it receives both a direct wave and a reflected wave that has reflected from the ionosphere. The mixing of these causes a phase shift that is unpredictable. This is referred to as intermodal interference.

Ground conductivity is the effect of a land mass affecting phase velocity, and is another predictable characteristic of VLF. The phase of the received signal depends upon the propagation velocity. This velocity depends upon whether the signal travels over land, sea, or ice. The signal travels fastest over water and slowest over ice.

Polar cap anomaly and solar flares are two unpredictable effects on phase velocity, but usually do not affect the ONS accuracy by more than 5 NM. Their effects are minimal and usually not long lasting. It has been noted that Omega does not perform well south of 80° latitude.

Correction factors can be programmed into more sophisticated Omega systems, thereby offsetting some of these negative effects.

QUESTIONS

1. How many Omega stations are there?
   a. 2
   b. 4

c. 6

d. 8

2. Omega operates in the _________ frequency spectrum.

a. LF

b. VLF

c. HF

d. VHF

3. An Omega station operates on:

a. four common frequencies and one unique frequency

b. three common frequencies and two unique frequencies

c. two common frequencies and two unique frequencies

d. one common frequency and four unique frequencies

4. The Omega system uses:

a. Lanes

b. highways

c. runways

d. skyways

5. How many stations must be received in the hyperbolic mode of operation?

a. one

b. two

c. three

d. four or more

CHAPTER 14

ARINC 429 Bus

Introduction

This chapter deals with the basic principles of the ARINC 429 busing system.

Objectives

Identify the term ARINC 429

Identify the purpose of the busing system

Explain how the data are transferred

Identify the frequencies of transmission

Identify the 32-bit format of the transmission

Explain logic high and logic low voltage levels

Explain general data link

Explain dedicated data link

Review the Binary number system

Review the Octal number system

Convert BCD to decimal

Basic Concept

The purpose of the ARINC 429 bus is to provide a method of data transfer that has increased accuracy, decreased weight, and decreased maintenance costs as compared to earlier aircraft system designs. This requirement comes about because current generation aircraft process large amounts of information transfer between systems and system elements.

BASIC CONCEPT

The purpose of the ARINC 429 bus is to provide a method of data transfer that has increased accuracy, decreased weight, and decreased maintenance costs as compared to earlier aircraft system designs. This requirement comes about because current generation aircraft process large amounts of information transfer between systems and system elements.

For example, in older generation aircraft the information exchange occurs on a system by system basis, that is, each system is separate and complete by itself and looks like Figure 14.1.

In newer generation aircraft the information is transferred on a bus system (see Figure 14.2). The **ARINC 429** is actually a specification that is published by *A*eronautical *R*adio *Inc.* which is a standards association that is endorsed by airlines, manufacturers, freight companies, etc. The ARINC 429 standard was published in December 1979, and is referred to as the Mark 33

Figure 14.1
Simplified Aircraft System Layout

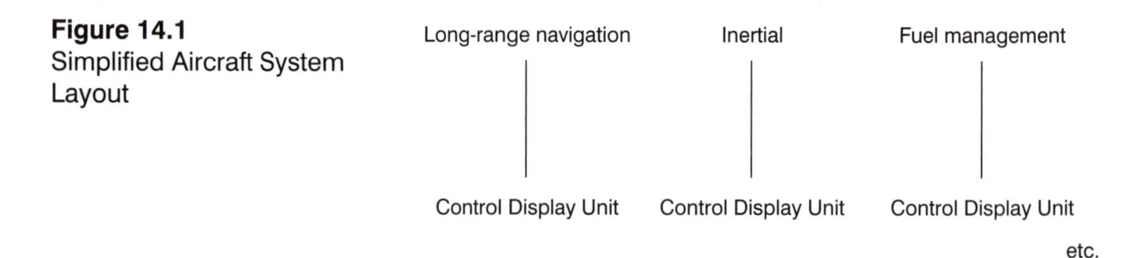

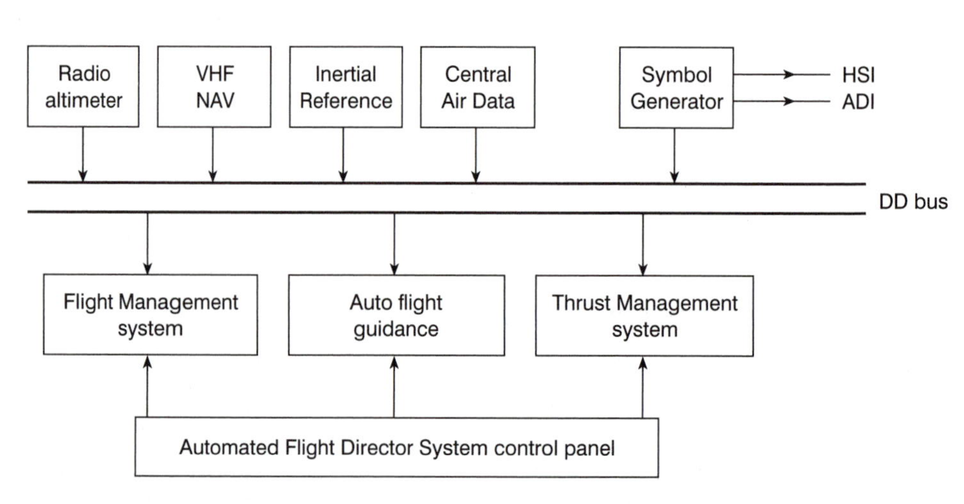

Figure 14.2
ARINC 429 Bus System

Digital Information Transfer System (DITS). Often a digital data bus is simply referred to as *DD bus.*

A *general data link,* as in Line 1, Figure 14.3, is a twisted pair of wires. Data are transferred from *A* terminals of box *X,* to *A* terminals of box *Y* and box *Z.* This general data link then provides the data that are available from box *X* terminal *A* to any number of other boxes in the system. Some boxes may require this information and some may not.

A *dedicated data link,* as in Line 2, Figure 14.3, is on a twisted pair, again in one direction only, and provides data from one specific location to only one other specific location.

ARINC 429 transfers data in one direction only down each data link. Other systems such as MIL 1553 can transfer data in both directions with the use of a bus controller.

Four Data Parameters of System

The digital transfer of information on the bus can represent four types of parameters. They are:

- *Discrete.* These parameters are actual voltage levels as either present or absent. For example, if the gear is up, the voltage is 28 V; when the gear is down, the voltage is 0.

Figure 14.3
Dedicated and General Data Links

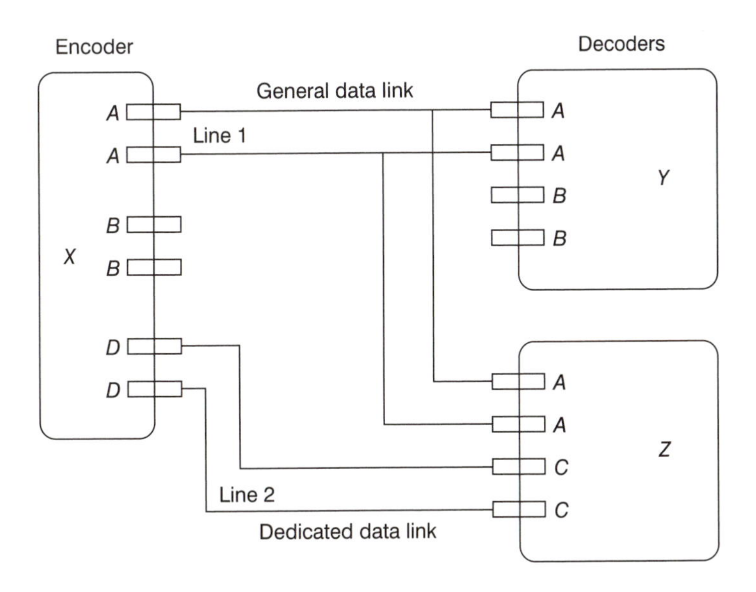

- *Numeric.* These parameters represent quantities—temperature, weight, pressure, etc.
- *Alphabetic.* Alphabetic parameters are letters, abbreviations, punctuation, etc.
- *Graphic.* Graphic parameters are symbols, lines, arcs, and vectors.

The 32-Bit DITS System

A 32-bit data word is used in the ARINC 429 DITS system and standard digital numbering systems are used in the creation of each word. The words are in **Binary** or in the **Binary Coded Decimal.**

- Bits 1 to 8 are called the *label.* The label is used to describe the system that the data relates to. There are only 256 code combinations with 8 bits and there is a requirement on the aircraft for more than 700 labels. One label can have more than one meaning. For example, the same code could be used for the #1 com and #2 MLS. Caution must be used in the wiring of the system so that these codes are not confused.
- Bits 9 and 10 are called the **SDI—Source/destination identifier.** The SDI is a steering code that sorts out which box is talking to which box.
- Bits 11 to 29 are called the *DATA,* sometimes referred to as the data field.
- Bits 30 and 31 are called the **SSM—Sign Status Matrix,** which identifies parameters such as direction, sign, value, north, south, etc.
- Bit 32 is called the *parity bit.* **Parity** checking is a method of testing for errors in the digital transmission. Figure 14.4 shows the format and labeling of the 32-bit word.

Figure 14.4
32-Bit Word Label

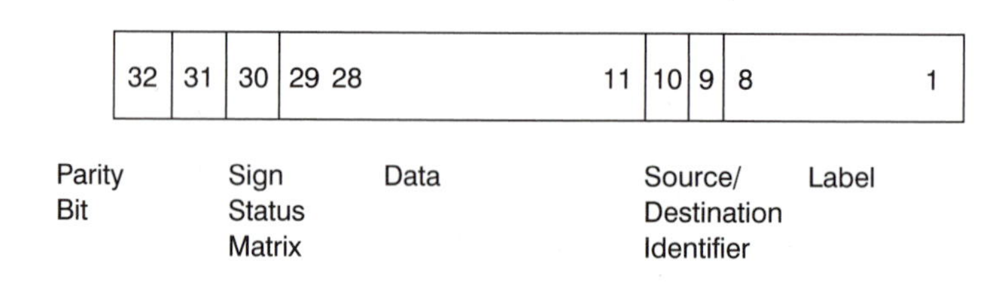

Data Transmission Rates and Voltages

Figure 14.5 shows an overview of the transmit side of the system. There are two rates of data transmission. The first rate is called the low rate and operates between 12 k and 14.5 k bits per sec. The high rate operates at 100 k bits per sec. The high and the low operating rates are not mixed on the same data bus.

The actual voltage level of the digital system is not 5 V like in TTL logic, or variable levels as in CMOS logic, but rather is set up for a differential +10 V for a logical one, and a differential –10 V for a logical zero, with the 0 V level being the "rest" state. The 429 uses an RTZ format (see Figure 14.6). RTZ means return to zero for each bit. This differential 10 V is created by one wire of the twisted pair going to +5 V while the other wire goes to –5 V. When this pulse is over, both lines return to zero V. The purpose of this format is to make the system more immune to noise. Any noise picked up by the bus will affect both wires the same and result in the differential voltage being 0, and thus the noise is ignored by the receiver.

The bus system uses a shielded twisted pair cable and is connected such that the shielding is grounded at each end of the cable. This assures a high degree of immunity to digital data distortion, as well as high voltage spike protection.

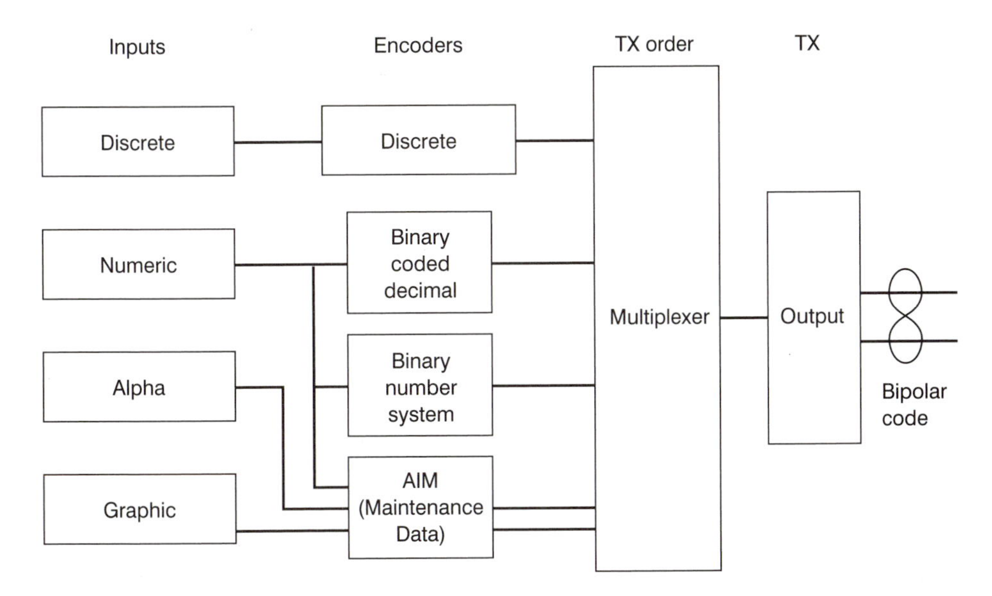

Figure 14.5
ARINC 429 Inputs

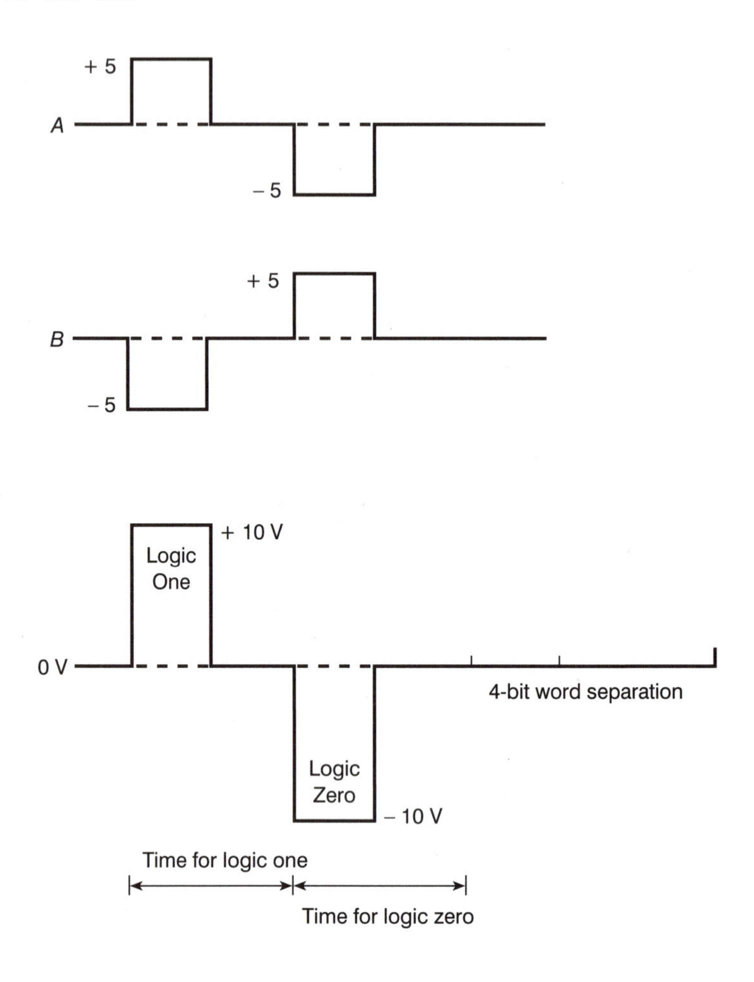

Figure 14.6
ARINC 429 Return to Zero Logic Format

The Receiver Outputs

Figure 14.7 shows what the receiver side of the system looks like. More details about the system follow:

- *Shift register.* A digital device to manipulate the serial data on the bus.
- *BCD channel.* **BCD** means **Binary Coded Decimal.** A number such as 097 can be represented as:

  0000 1001 0111

  The BCD words are used for ground speed, wind speed, selected course, and tire loading.

Figure 14.7
ARINC 429 Outputs

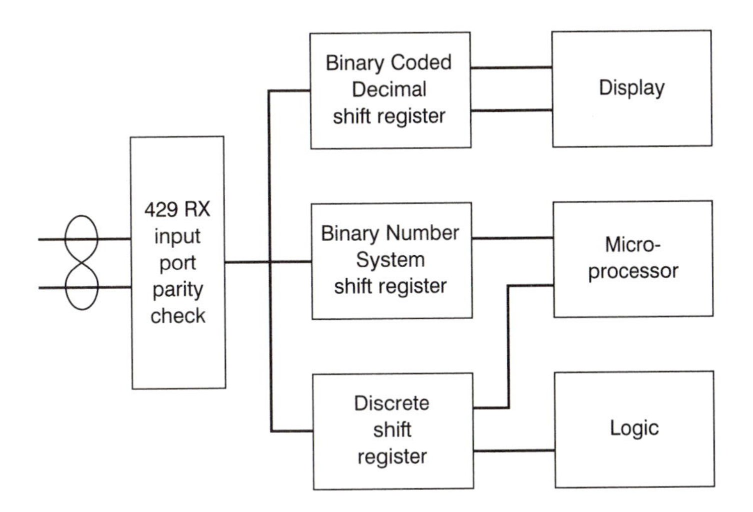

- *BNR channel.* **BNR** means *binary number.* The same number (097) can be represented as a standard format Binary number that uses zeros and ones in a base two numbering system.

Weight	64	32	16	8	4	2	1
Code	1	1	0	0	0	0	1

- *Discrete channel.* The discrete channel decodes the multiplexed signal from the transmitter to check for the presence or absence of a voltage.

QUESTIONS

1. Name the four types of data parameters in the bus system.
2. Identify characteristic elements of alphabetic.
3. What is the low data transmission rate?
4. What is the logical high voltage level?
5. What is the logical low voltage level?
6. What is the voltage level during the time when no data are being transferred?
7. What is odd parity?
8. What type of wiring is used for the 429 bus?
9. What does the abbreviation DITS stand for?

CHAPTER 15

Global Positioning System

GPS

Introduction

This chapter deals with the basic principles of the Global Positioning System of navigation.

Objectives

Identify the purpose of the GPS

Identify the three basic parts of the GPS

Tell how many satellites are in the system

Tell how many ground monitoring stations there are around the world

Tell where the master control station is located

Explain the functional operation of one-, two-, and multi-channel receivers

Explain how long it takes for a satellite to complete one orbit

Identify the frequencies of operation

Discuss the accuracy of the system

Basic Concept

Using transmissions from orbiting satellites it is possible to establish a very accurate three-dimensional navigation fix.

The Global Positioning System was conceptualized near the end of the Second World War. The current **Global Positioning System (GPS)** program was started in 1973 with the joining of two existing experimental satellite navigation programs: the US Navy's Timation Program and the US Air Force's 621B Project. The culmination of the two projects under the US Air Force resulted in the current GPS program. GPS was initially developed for military application to provide a worldwide three-dimensional information system of:

latitude
longitude
altitude
time and speed

The GPS application was intended for precision-aimed weapons and accurate troop deployment. GPS was first successfully tested and proved in the Gulf War.

The financial support for the development of the system was from taxpayers, and under political pressure the US Congress was forced to pressure the military to make GPS available for civilian use. GPS is now widely available for almost any application, from navigation of all types of vehicles to use by recreational hikers and in seismic and surveying applications. GPS is a rapidly developing 24-hour navigation system with almost unlimited uses. For civilian use, it can tell you where you are on the earth's surface or atmosphere within 30 m, and how fast you are going within 0.1 mps. It can also tell you the time within 1 sec in 70,000 years. Military specifications are even more accurate, but are not available for civilian use. The main advantages of GPS are:

accuracy
worldwide application
signals that are not contaminated by the weather

The major disadvantage of GPS is that the system relies entirely on the fidelity of the satellites' data transmission. The US military is in charge of the satellites.

GPS BASICS

The GPS theory of operation is relatively simple. Orbiting satellites transmit coded pulses that contain the satellite's position and the precise time of the transmission. The GPS receiver measures the elapsed time and then converts this to distance. By measuring the distances from three satellites, exact three-dimensional positioning is available. The problem of exact timing would normally be solved by having time-correlated atomic clocks on all satellites and on all GPS ground receivers. This would make the GPS receiver very expensive—in the order of $100,000. To solve this, a fourth satellite is used to provide precise timing for calibration of the ground receiver's clock. This reduces the receiver cost to $3,000–$5,000. There are nonaviation application GPS receivers that sell for less than this. For three-dimensional navigation, four satellites are needed. For two-dimensional navigation, three satellites are needed.

As indicated in Figure 15.1, there are three basic parts of the Global Positioning System:

the space segment
the control segment
the user segment

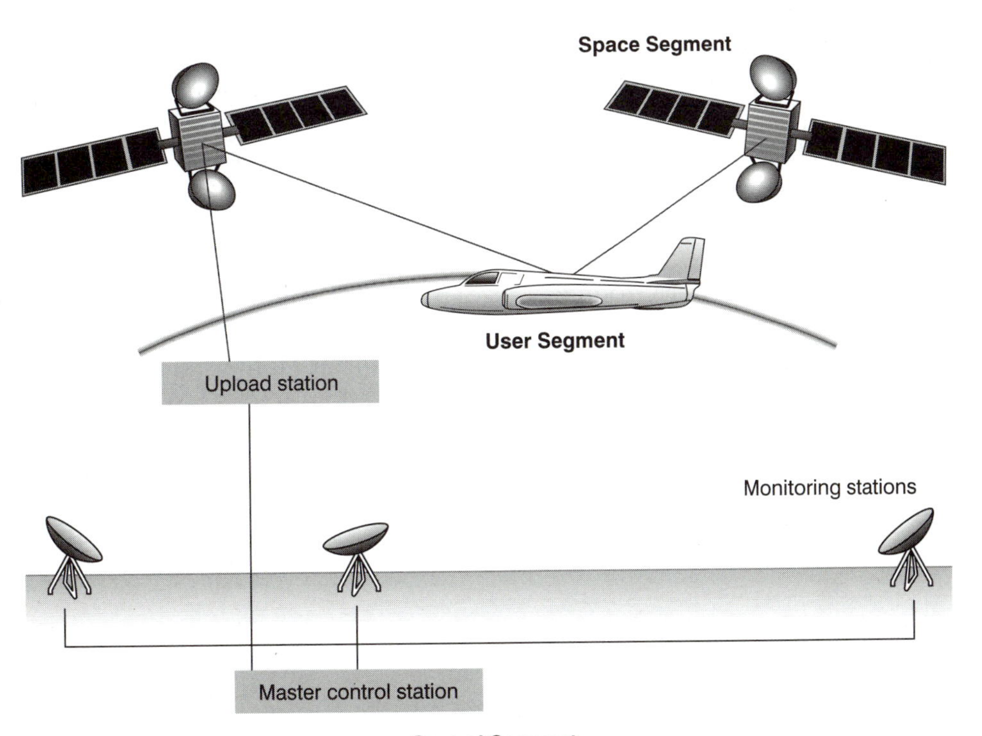

Figure 15.1
GPS System Segments

The Space Segment

The space segment consists of 21 operations satellites and three active spares. The satellites continuously transmit an encoded signal containing all necessary information for system operation. The receiver decodes the signal and processes the data using the time delay between signal transmission and signal reception. These calculations can determine the speed, position, and altitude of the receiver. The system assumes that the satellite clock and the receiver clock are perfectly synchronized. In practice this is virtually impossible, but with use of four satellites the receiver can compute the difference in the clocks by solving for four unknown variables. Two-dimensional (Two D) operation uses three satellites and a known altitude to obtain the four measurements needed to calculate position. Two D is typically used for ground or sea operation where the altitude is known.

The Control Segment

The control segment consists of a master control station in Colorado Springs, Colorado, and five linked monitoring stations around the world. These stations are located at Hawaii, Ascension Island (South Atlantic), Diego Garcia Island (Indian Ocean), Kwajalein (South Pacific), and Cape Canaveral (Florida). The control segment is responsible for monitoring all of the satellites and ensuring that they are all operational. The control segment has a Master Ground Control Station in Colorado Springs with the backup master control at Onizuka Air Force Base in California. The master control station can uplink through the monitoring stations, and on a 24-hour-a-day basis make any specific satellite corrections to orbit, speed, etc., as required. The satellites are tracked from the monitoring stations and in this way the master control station can provide correctional data three times daily to every satellite.

The User Segment

The receiver is strictly a passive device. It can only receive and process the information transmitted by the satellites. There are three basic systems used to receive and process satellites' transmitted data. These are referred to as one-, two-, and multi-channel GPS receivers.

The one-channel system is the simplest. It can track and receive information from only one satellite at a time. It samples the information from the four required satellites sequentially and then processes the information. The one-channel system requires 12.5 min to download the satellite's position information, so that the receiver can choose the four satellites that will give the best positional accuracy. It also requires 30 sec per satellite to download navigational data from the four selected satellites. Single-channel receivers are therefore useful only for slow-moving applications. (Fast-sequencing single-channel receivers are available, but accurate data cannot be guaranteed due to inherent noise factors from the devices and from the environment.)

The two-channel receiver is twice as fast as the single-channel one in downloading satellite position information. One channel can be devoted to monitoring all satellite position data while the other channel continues to sequence through the four satellites for navigational data. However, even with two-channel operation these receivers are useful only for relatively low velocity applications such as light aircraft, ground vehicles, ships, etc.

Multi-channel receivers are required for high-speed applications. Ideally they have five channels—four channels to provide continuous position information and velocity updates needed for navigation. One channel is dedicated to select the four best satellites from the downloaded satellite position information.

THE SATELLITES

The satellites are in six orbital planes with four satellites in each plane. It takes the satellites 12 hr to complete one orbit. At any given time any one satellite has a 28° view of the earth. The earthbound GPS receiver should be able to see six satellites at any time of day. With the downlinked positioning information of the satellite, the receiver can take advantage of choosing from which four satellites to obtain the navigational information that will provide the most accurate indication. Relative location of the satellites visible to the GPS receiver is known as satellite geometry. The more separation there is between the four needed satellites, the greater is the accuracy for calculating navigational information.

Each satellite weighs 1,016 lb and is at an altitude of 10,898 NM. On board are four precision atomic clocks. Power is supplied by nickel cadmium batteries and solar panels. Each satellite is designed with a life expectancy of 7.5 yr, but early test satellites have long outlived this.

THE SIGNAL

Each satellite transmits two different L Band signals, each with a 20 MHz band width. These are referred to as signals L1 and L2. L1 is centered at 1575.42 MHz, and L2 at 1227.6 MHz. The two frequencies are needed to correct for ionospheric propagation delay, made worse during times of high solar activity such as solar flares and sun spots. A single frequency receiver will have an inherent error of 25 m or more depending on solar activity. L1 and L2 are biphase modulated. **Pseudo Random Noise (PRN)** also causes the relative phase to vary in a predictable manner. This makes better use of the frequency spectrum. Each of the 24 satellites has its own distinct PRN, which allows the GPS receiver to distinguish between them and decode the biphase modulation.

Two types of codes are transmitted:

Coarse/Acquisition (C/A) code and

Precision (P) code

The P code is 10 times faster than the C/A code but does not necessarily have 10 times greater resolution. The P code is for use by the US Department of Defense and other authorized users. The C/A code is for civilian use and therefore available to anyone. It has an accuracy of approximately 30 m. This was considered by the US military to be too advantageous to an adversary during a time of war. Since there is no control over who uses the GPS, the military has designed in the ability to manipulate the accuracy of the transmitted data, called **Selective Availability (S/A).** This reduces the accuracy of the C/A code to 100 m, and the P code accuracy from 10 m to 100 m. S/A can be overridden only by authorized users determined by the US Department of Defense and the US government. To protect against false transmission by a foe during times of war the US military has a secret Y code that is transmitted only at the discretion of the military. It is comparable to the P code.

		SPS *(Unauthorized)*	**PPS** *(Authorized)*
Selective ability	C/A code	> 100 m	~30 m
	P code	> 100 m	~10 m
Antispoofing	Y code	No access	~10 m

Present store-bought GPS receivers come with warnings and have a disclaimer stating that they are not to be used for navigation because of the

S/A. For example, the Sony IPS 360 handheld GPS owner's manual states, "Warning: GPS system signals are controlled, maintained, and operated by the Department of Defense of the United States. Without notice the DoD can change the characteristics of the signal, which will degrade the accuracy of this unit. Use the Sony IPS 360 at your own risk." This particular unit advertises an accuracy of 30 to 100 m positional and 0.3 knot vectorially with maximum operating speed of 530 knots.

There are at least seven manufacturers of panel-mounted and portable-mounted GPS receivers for general aviation use. Companies like ARNAV, Allied Signal General Aviation Avionics, Garmin, Magellan, II Northstar, II Morrow, and Trimble are developing receivers, as well as a host of others who are considering getting into the game. Prices range from $2,000 to $10,000 and the list of options and interfaces has never been so great for the avionics consumer, nor has the competition been so great.

GPS OPERATING DATA

GPS operating data are:

Code **(PRN)**	*Chip Rate*	*Band Width*	*Signal Strength*
P code	10.23 MHz	20 MHz	160 dBW
C/A code	1.023 MHz	2 MHz	160 dBW

GPS displays the latitude, longitude, and altitude determined by the world geodetic system, which is a coordinate system based on defining the whole earth as an ellipsoid that revolves on its axis. Conventional world atlases may use a different geodetic system and therefore the latitude and longitude displayed by a GPS receiver may not be identical with that found in an atlas. GPS receivers may indicate either true bearing or magnetic bearing.

GENERIC GPS RECEIVER

As seen in Figure 15.2, the basic receiver consists of:

- *Antenna and the preamplifier.* Antennas used for GPS receivers have broad beam characteristic, thus they do not have to be pointed to the signal source like satellite TV dishes.

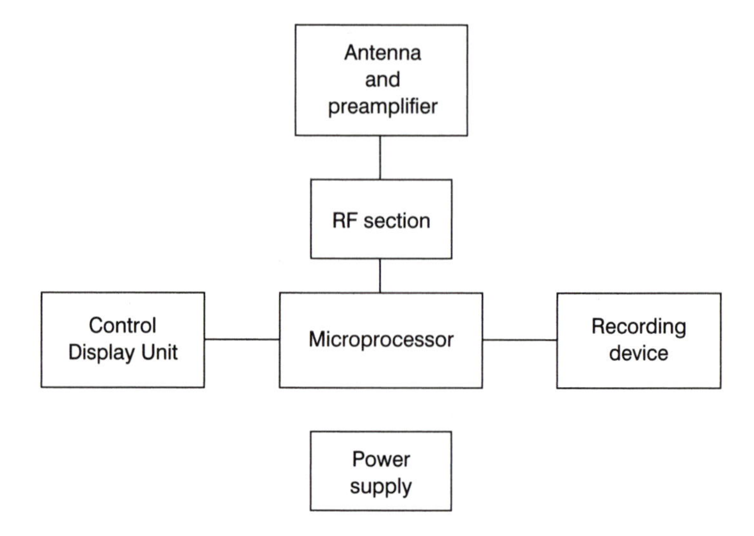

Figure 15.2
Typical GPS Receiver

- *RF section.* The RF section contains the signal processing electronics in a combination of digital and analog circuits. Different receivers use different techniques to process the signals. They are:

  code correlation

  code phase and frequency

  carrier signal squaring

  The RF section consists of channels using one of these three approaches to track a received GPS signal. The number of channels varies between 1 and 7 for various manufacturers.

- *Control display unit.* The control display unit enables the operator to interact with the microprocessor. Its size and type varies greatly for different receivers and applications, ranging from handheld to a video monitor with full-size key pad.

- *Recording device* (two options). Either tape recorders or floppy disks are used to record the observations and other useful information extracted from the received signal.

- *Power supply.* Receivers need only a reliable low voltage DC power supply.

A more detailed circuit diagram of a dual frequency multi-channel C/A and P code receiver is shown in Figure 15.3.

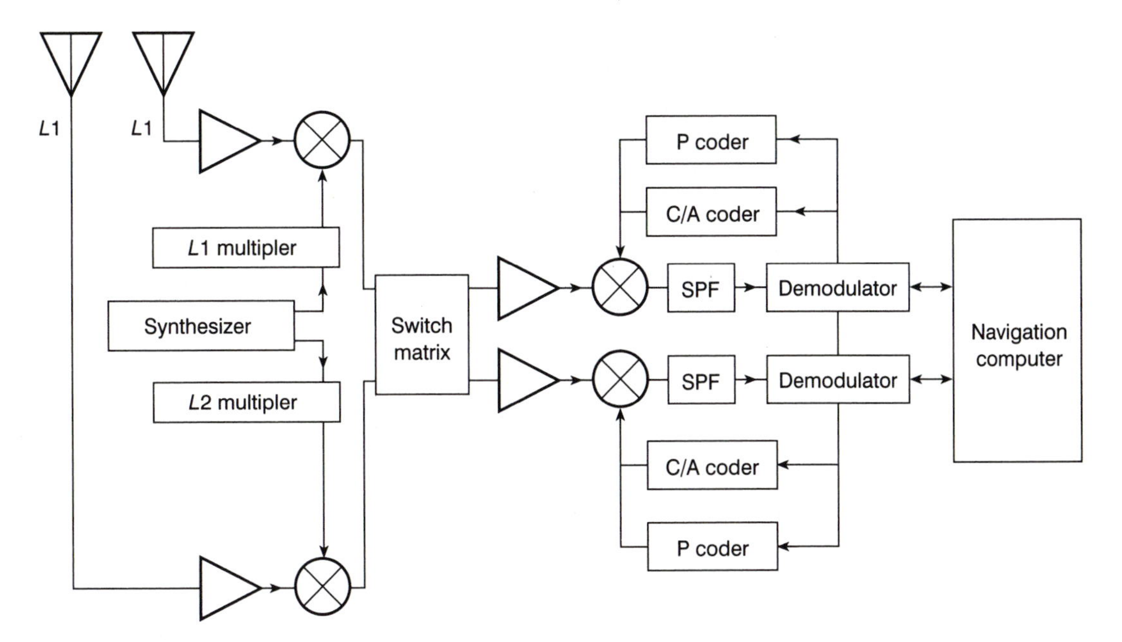

Figure 15.3
Typical GPS Receiver—Detailed Block Diagram

GPS SATELLITES

There are three generations of GPS satellites, the Block 1, Block 2, and the Block 3 units. Basic functions of the satellites are to:

- receive and store information transmitted by the control segment
- perform limited data processing by means of onboard microprocessor
- maintain a very accurate time base through the use of 4 oscillators, 2 cesium clocks, and 2 rubidium clocks
- transmit information
- execute satellite maneuvering by means of ground-controlled thrusters

Solar panels that are 7.23 m^2 are used to charge the batteries. The batteries supply power when the satellite is in the earth's shadow.

As of 1992 the USAF had 13 Block 2 and 4 Block 1 operational satellites. At that time these 17 satellites were providing two-dimensional and three-dimensional navigation coverage for most of the world and for most of the 24-hour day. This coverage and reliability is often referred to as "confidence level." So we could say that the confidence level is 80%.

As of June 1993 there were 23 satellites in the constellation, thus increasing the coverage and time to almost complete. The satellite constellation was complete by December 1993, and now has 18 operational satellites plus 3 active spares in 6 orbital planes. Each satellite takes 12 hr to complete an orbit. This provides 24-hour-a-day coverage anywhere in the world. Finally, on July 17, 1995, the system officially reached full operational capability.

Initial Operational Capability (IOC) Problems

The constellation of satellites was complete in late 1993 and marked by a declaration of **Initial Operational Capability (IOC).** The IOC was marked by temporary gaps in operational capability of the system due to testing and calibration, and during this phase was not 100% reliable.

At this juncture one would think that GPS is the end all to the navigational problems. Wrong. The political game and the implementation are creating some problems.

Remember that the US military developed the system and was forced by the US Congress to provide a civilian component to the system. In this compromise the US military maintains the control of the system and is able to downgrade the accuracy of the system at any time. In a time of national emergency the DOD can command the satellites to "dither" their signal accuracy. This will result in all ground receivers having large errors of more than 1,000 m. The US military has specially equipped GPS receivers that contain the antispoofing digital antidote. This condition is of prime concern to civilian aviators.

The system accuracy is nominal at 15 m. This accuracy is a function of atmospheric signal, receiver accuracy, satellite positional accuracy, and updated ephemeris data. Some receivers are designed to account for some of these parameters and have much higher accuracy.

The DOD generally believes that 15 m is too accurate for a potential foe, and therefore introduces error levels to about 100 m. This error is acceptable for enroute IFR navigation solutions but is not acceptable for precision approaches or zero/zero taxi navigation (landing in weather of zero visibility).

The FAA is in the certification process and working of TSO C129. This certification will provide for the use of GPS as a supplemental IFR enroute navigation system. Transport Canada is following closely on the heels of the FAA. In a Special Aviation Notice dated July 22, 1993, Transport Canada provides conditional approval for GPS in certain IFR applications, keeping pace with US Federal Air Regulation (FAR) standards.

The goal of Transport Canada (TC) is that users obtain benefits from GPS without sacrificing safety and TC is therefore approving GPS usage in stages.

The Before IOC Stage in Canada

Currently, GPS can be used as primary IFR flight guidance for oceanic, domestic enroute, and terminal flights if:

- the GPS receiver has TSO C129 approval
- the installation is according to the airworthiness manual

The After IOC Stage in Canada

GPS will not meet stringent integrity and availability requirements of IFR operation. (By definition, integrity is the ability of a navigation aid to notify a pilot if it is radiating erroneous signals. Example: VOR shuts down automatically if out-of-tolerance conditions are detected. GPS can take hours to notify users of unhealthy conditions.)

The integrity problem is being explored by:

- using ground monitor stations
- using TSO 129 for IFR GPS receiver calls for **Receiver Autonomous Integrity Monitoring (RAIM).** GPS positions require 4 satellites and RAIM determines unhealthy satellite and stops using it and informs the flight crew. RAIM requires 6 satellites in view to process the combinations required to determine an unhealthy satellite. Unfortunately, 6 satellites are not always in view.
- using traditional navigation aids (VOR and NDB) to back up GPS as an interim solution
- monitoring GPS unless RAIM-equipped
- having the flight crew revert to operating with the traditional NAV aids if there is any discrepancy between GPS and the traditional NAV aids

Two modifications to the GPS system are:

- *Wide Area Differential GPS (WADGPS).* Differential GPS (DGPS) consists of ground-based stations that collect and calculate position corrections, then use a VHF data link to airborne traffic. June 1993 saw Wittman Regional Airport as the only one equipped with DGPS. The VHF data uplink typically is at 169.375 MHz.

- *Ground Integrity Broadcast (GIB).* GIB will provide the minimum 6-sec requirement satellite failure warning.

Neat things are also being done with GPS over and above the latitude, longitude, altitude, air speed, and ground speed. Some of these are:

- attitude—in three axes to 1/10 of a degree
- wing flex—1.4 mm vertical acceleration
- windshear—microburst downdraft warning, about 20-sec advance notice
- automatic dependent surveillance (ADS)—GPS position and velocity data linked to ATC are better than provided by radar; there is also a TCAS link.

GLONASS

The Russians started developing their ***G**lobal **O**rbiting **N**avigation **S**atellite **S**ystem* **(GLONASS)** in 1982. Their system is almost identical to GPS with notable differences:

They have 24 satellites in three orbital planes.
Distance is 10,300 NM.
Orbital period is 11 hr 15 min.
Orbital inclination is 65°.
There are two frequencies of:
F1 at 1597 MHz to 1617 MHz band
F2 at 1240 MHz to 1260 MHz band.
Selective ability is not available on GLONASS.

In April 1993 the Soviets had 61 satellites, of which only 15 were operational.

In 1988 at an International Civil Aviation Organization (ICAO) meeting there was apparent interest in developing a combined GLONASS/GPS application, or GNSS. This would be advantageous as it would radically increase the number of satellites and therefore the confidence level. Currently, however, the Russian satellites are not performing well and newly launched satellites are not expected to perform as well as American ones.

Another major problem to an integrated GNSS is that the Russians used the 1985 Soviet Geocentric System for expressing the coordinate system while the Americans used the 1984 World Geodetic System. Both use different time coordinate systems and this problem also must be resolved before a joint integrated GNSS can be developed.

GPS TECHNICAL DATA

Satellite Geometry

Figure 15.4 shows GPS satellite geometry. In GPS, all 6 orbital planes are inclined by 55° to the equatorial plane and rotated by 60° with respect to their neighboring planes. Thus their separation in right ascension is 60°. The 3 active spares numbers (19, 20, 21) are in orbital planes labelled 1, 5, and 3, and have arguments of latitude of 30°, 310°, and 170° respectively at the instant when satellite 1 ascends through the equatorial plane.

The orbit is circular with a nominal altitude of 20,183 km and the orbital period is 12 sidereal hr (half of the earth's period of rotation).

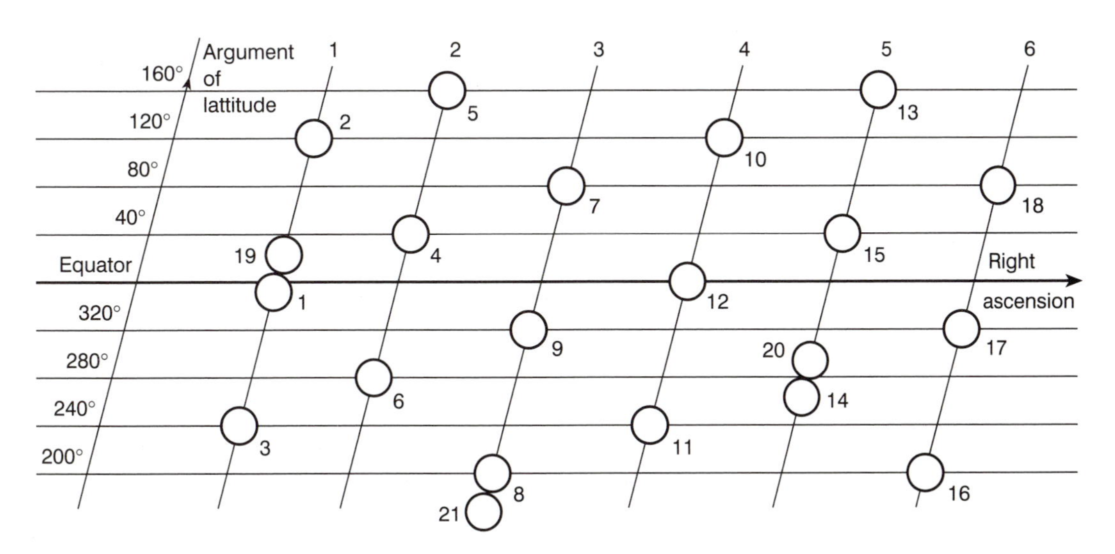

Figure 15.4
GPS Satellite Geometry

One-Way Ranging

It is practically impossible to keep the transmitter clock and receiver clock perfectly synchronized. For example, a 1 μs error equals a 300 m error in calculated distance between the transmitter and receiver. The solution to this problem lies in recreating a replica code in the GPS receiver, as indicated in Figure 15.5. Each clock will run at its own rate, keeping its own time. However, if the relationship between the two time bases of the clocks is known then they still can be said to be synchronized.

The GPS control segment monitors all the satellite clock differences and includes this information in the satellite message broadcast to the user; thus all satellite clocks can be considered synchronized. The receiver clock is then corrected by measuring the time shift required to line up a replica of the code generated in the receiver with the received code from the satellite.

Knowing the time it takes for the signal to travel from the satellite to the receiver will allow calculation of the distance to the satellite. This value must be corrected for distortion by the atmosphere and other factors and is therefore referred to as "Pseudo Range." It is derived by aligning a receiver-generated replica of the arriving signal with the actual arriving signal from the satellite. If the ranges to four satellites are combined with the orbital description, then the fourth range is used to account for the behavior of the receiver's clock. The receiver can then determine its three-dimensional coordinates.

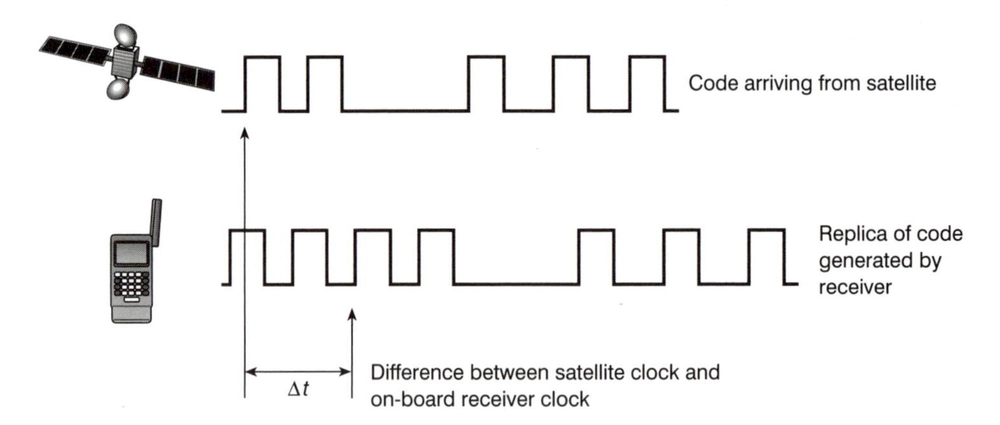

Figure 15.5
GPS One-Way Ranging

GPS Signal Structure

Every satellite transmits two frequencies for positioning: 1575.42 MHz and 1227.60 MHz. The two carriers (L1 and L2) are coherent and modulated by various signals.

The first code is known as the C/A code and consists of a sequence of logic ones and logic zeros, emitted at the frequency F1/10 = 1.023 MHz which repeats itself every millisecond. The second code, known as the P code, consists of another sequence of ones and zeros, emitted at a frequency of Fo = 10.23 MHz and repeats itself every 267 days. The Y code can be used instead of the P code; however, the equation that generates the Y code is classified by DOD, thus rendering the system useless to anyone who does not have access to it.

The concept of modulation is as follows: if a bit value is one, the carrier phase is shifted by 180°, and if the bit value is zero there is no change in the carrier. This is referred to as *Differential Phase Shift Keying* or *biphase shift keying,* shown in Figure 15.6.

Both carriers are modulated, with the broadcast satellite message being a low frequency 50 Hz stream of data designed to inform the user about the health and position of the satellite.

Message Format

The GPS message is formatted into frames of 1500 bits. At the 50 bits per sec message data transmission rate, this takes 30 sec to transmit. Each frame contains 5 subframes, and each subframe contains ten 30-bit words. For sub-

Figure 15.6
Differential Phase Shift Keying

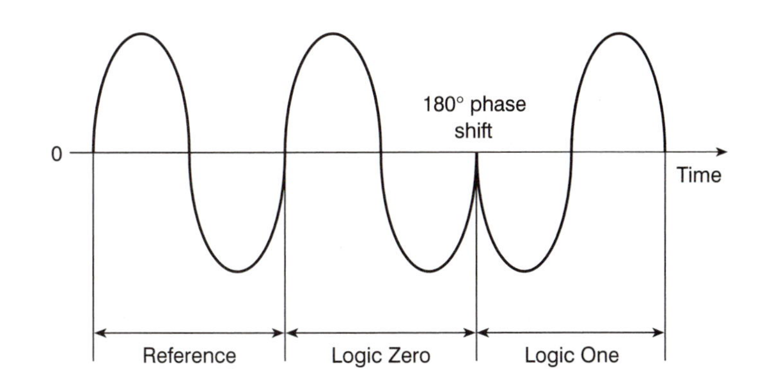

frames 1 to 3, the information does not, in general, change from frame to frame. However, for subframes 4 and 5, consecutive frames contain different pages. There is a total of 25 such pages for each of the 4 and 5 subframes. One master frame contains all the information in all pages of subframes 4 and 5 and consists of 25 complete frames. One master frame takes 12.5 min to transmit. This system is shown in Figure 15.7.

Each message bit lasts for 20 ms. During this period the complete C/A code will have repeated exactly 20 times. A divide by 20 counter on the C/A code output provides the timing for the message bits.

Message Content

The five subframes contain the following data:

Subframe 1 contains satellite clock correction coefficients, various flags, and the age of the data.

Subframes 2 and 3 contain the broadcast ephemeris parameters.

Subframe 4 contains an ionospheric model, UTC data, flags for each satellite indicating whether antispoofing is on, and the almanac data and health for satellites 25 and up if more than 24 satellites are in orbit.

Subframe 5 contains the almanac date and health status for the first 24 satellites.

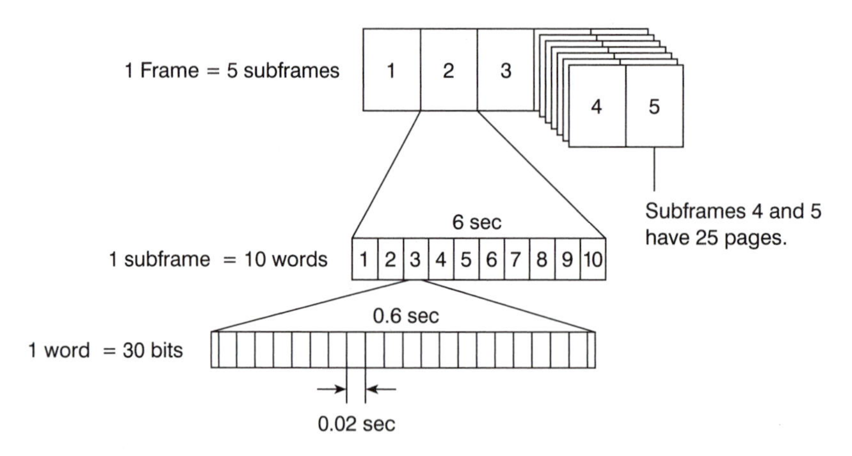

Figure 15.7
GPS Message—Frames

QUESTIONS

1. Identify the purpose of the GPS and the basic principle of operation.
2. Identify the three basic segments of the GPS. Comment briefly on specifics of each segment.
3. How many satellites are in the constellation?
4. How many ground monitoring stations are there around the world?
5. Where is the master control station located? The backup master? The other monitoring stations? What do they provide?
6. Explain the functional operation of one-, two-, and multi-channel receivers.
7. How long does it take for a satellite to complete one orbit? At what height?
8. How many satellites are in one orbit?
9. What are the two frequencies of operation?
10. What is the accuracy of the system?
11. What is GLONASS?
12. As of 1995 the GPS constellation was complete. How many hours per day is this coverage available?
13. How many satellites are required for a high accuracy three-dimensional navigation fix?
14. How many satellites are required for a high accuracy two-dimensional navigation fix?
15. How many satellites are required for time solutions?
16. What are the two major weaknesses of the GPS?
17. About how long does it take to initialize a GPS receiver when first acquiring the satellite signals?
18. How long does it take to update a complete set of data in the receiver?

CHAPTER 16

Flight Control Systems

Autopilot

Introduction

This chapter deals with the basic principles of autopilot systems.

Objectives

Identify the purpose of a flight control system

Describe the operation of a basic control loop

Relate the basic control loop to the autopilot

Relate the basic control loop to the pitch channel

Relate the basic control loop to the roll channel

Relate the basic control loop to the yaw damper

Basic Concept

The primary task of the autopilot is to measure the position of the aircraft relative to the horizon, and maintain that position by deflecting the appropriate control surface.

An **autopilot** is a system that employs position control. Figure 16.1 shows the basic position control system. Position control requires the movement of a load. This movement is normally accomplished with an electric motor as an actuator connected to the load with mechanical linkage. The electronic system that controls the motor is usually referred to as a servo loop.

A **servo loop** contains a position-sensing device that measures the present position of the load. This position signal is sent to the summing amplifier, which compares this signal to the desired position of the load (command signal) and amplifies the difference (error signal) in order to provide motor drive.

The mechanics of the motor and gear reduction will cause the load to overshoot its commanded position, which will cause an oscillation. To reduce this effect, rate feedback is often employed. Rate feedback produces a signal relative to the velocity of the motor and of a polarity that will tend to cancel the error signal.

The autopilot portion of a flight control system is a simple system of positioning control. Its primary task is to measure the position of the aircraft relative to the horizon and maintain that position by deflecting the appropriate control surface. The term **Flight Control System (FCS)** is generally used to indicate the combination of an autopilot system and a **Flight Director System (FDS).**

Very commonly the Flight Director System uses a **Flight Director Computer (FDC)** in conjunction with an **Air Data Computer (ADC).** The ADC acts as a primary input to the FDC. The autopilot uses the driving signals

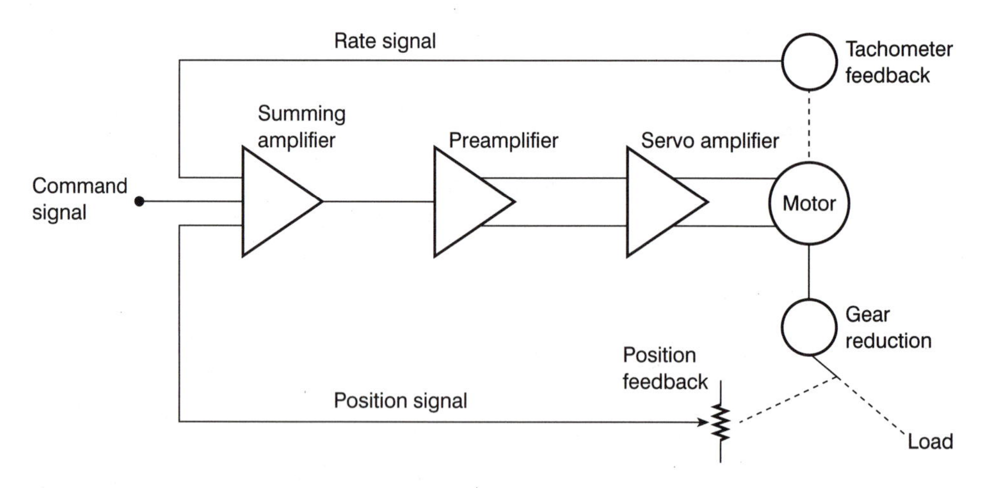

Figure 16.1
Basic Position Control System

from the FDC to position the aircraft in a navigational and attitudinal location desired by the pilot. Very simple autopilots have a limited number of input options to control the aircraft.

The less sophisticated the flight control system, the greater the amount of pilot monitoring required. All FCS systems require that the pilot monitor the performance of the aircraft! The simplest type of autopilot is a single axis autopilot or wing leveler. This system controls the roll axis of the aircraft only.

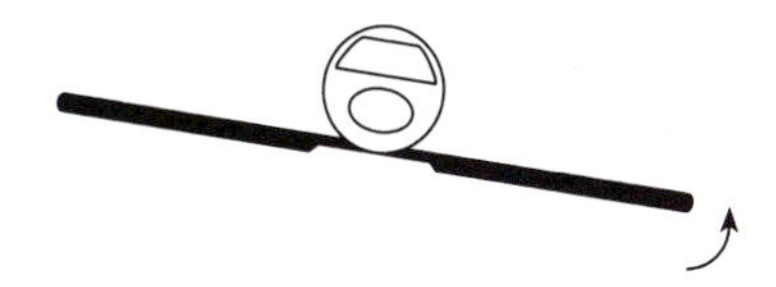

The pitch attitude of the aircraft must be flown manually. The pitch axis of the aircraft is different than the roll axis, because unlike the roll axis, the position of the aircraft in the pitch axis is required to change due to many factors.

In the following example, a small amount of control deflection is assumed. For example, if the position of the ailerons was deflected a certain amount, and this were maintained, the aircraft would roll in the appropriate direction until the amount of tendency to roll away from level flight was cancelled by the tendency of the dihedral to return the aircraft to level flight. The aircraft would maintain a turn in this direction that would stabilize to a given rate of turn. The other parameters of the aircraft would have very little effect on this action. In contrast to this, if the elevators of the aircraft were deflected down, the aircraft would pitch down, and the aircraft would lose altitude. This would also cause an increase in airspeed. This increase in airspeed would cause greater lift. This greater lift would slow down the loss of altitude and the aircraft would stabilize at another airspeed, but once again in level flight. Pitch trim is another input to the pitch system.

The third axis of the aircraft is controlled by the yaw damper system. The yaw damper is an independent control system that may or may not be a part of the autopilot system.

A more advanced autopilot than the wing leveler is the two-axis autopilot. This system controls both roll and pitch. A two-axis autopilot is more than twice as complex as the wing leveler because the pitch servo that controls the elevator would very quickly wear out if it were not for the pitch trim system that relieves the pressure on the controls. If a certain pitch attitude is desired and the elevators are positional to attain this, the aircraft will stabilize in this new position; however, there will be a constant force on the elevator required to maintain this condition. The trim tab is then adjusted to relieve the pressure on the control surface. The pitch trim circuit measures the tension on the control cables and electrically trims the aircraft until the tension is reduced.

Regardless of the level of sophistication of the autopilot there are certain common points to all of them. To keep the airframe safe from overstress, the pitch and roll parameters must be kept to values within an absolute limit. Most autopilot systems will not allow greater than a rate one turn, or a climb or descent angle of greater than 10°. To have the aircraft exceed these limits is very dangerous and could result in an accident.

One type of autopilot is commonly STC'd for more than one type of aircraft. For example, the KFC 200 is a light aircraft general aviation autopilot that is approved for many aircraft. Because the amount of control deflection and the speed of control deflection is different for each aircraft, the autopilot must be programmed for each type of airframe. If by chance, the wrong "flavor" of autopilot computer were to be installed in an aircraft, the results could be disastrous—overstress of the airframe and possibly crash.

All autopilots must measure the attitude of the aircraft in some way.

Some aircraft (the less expensive ones) measure the pitch attitude by monitoring the airspeed of the aircraft. If the aircraft were to pitch down, the airspeed would increase and the autopilot would deflect the nose up in order to correct the error. Pitch channel on the more sophisticated units measures pitch attitude with a gyroscope

The gyroscope is a spinning mass that maintains its orientation in space. The spinning mass is held in position by a framework called a gimbal. A gyroscope is a device that creates a stable platform in space by spinning a wheel-shaped mass at very high RPM. This mass is mounted on gimbals. The spinning mass is held in place by gyroscopic action and remains either vertical or horizontal in space depending on the function of the unit. The artificial horizon instrument uses a gyroscope with its spinning axis in the vertical direction, and therefore is called a vertical gyro. The stabilized directional compass system uses a gyroscope with its spinning axis in the horizontal direction, and is called a directional gyro. As the aircraft flies, its attitude is measured by comparing the position of the spinning mass with the position of the gimbal by the use of a synchro pickup. The synchro pickup is like a transformer with one primary (rotor) and several secondaries (stator windings) located at different angles. By comparing the phase and voltage of the secondary signals with each other, the angle between the rotor and stator can be measured.

This pitch measurement signal is then summed with the command input signal, and on a few systems with the position feedback potentiometer on the servo. The Cessna 400 system, for example, uses a potentiometer to close the feedback loop, whereas the King KFC 200 uses a rate feedback. Both autopilots use trim to relieve the pitch servo effort.

The King unit actually measures tension on the capstan to determine if trim is required. Other units simply measure drive to the pitch servo and average the requirement for servo operation. If a pitch command is called for that lasts for longer than 4 to 5 sec, a pitch trim signal will run the trim system to relieve the effort.

A similar situation exists for autopilot roll systems. The less expensive units don't measure a gyro signal, but rather rely on a roll error signal causing a change in heading. The roll limiting therefore depends on the rate of heading change. The more expensive and sophisticated units have a roll gyro output (same as pitch).

The aircraft servo motors are attached to the control surfaces by clamping onto the main control cable with a bridle cable and capstan, as shown in Figure 16.2.

The cable tension of the main and bridle cable is important for proper autopilot operation. The trim servo works differently due to the fact that the trim control cable is thinner than a main control cable. In this case, an extension is added to the trim cable at the turnbuckle, and the cable itself is wrapped around a continuous capstan. The servo mechanism must release from the primary controls using a clutch. It is very important that this function is working correctly or a situation could occur where the control of the aircraft is lost. A second level of fail-safe is added as a slip clutch so that if a malfunction occurs and the autopilot *cannot* be disengaged, the pilot can overpower the servo motor until the power to the autopilot can be removed. The King KFC 200 has four methods of disengaging the autopilot:

- disengage button on the control wheel
- engage the aircraft electric trim (manual)
- engage lever in the KFC 200 Mode controller
- pull the circuit breaker.

Figure 16.2
Motor, Capstan, and Bridle Cable

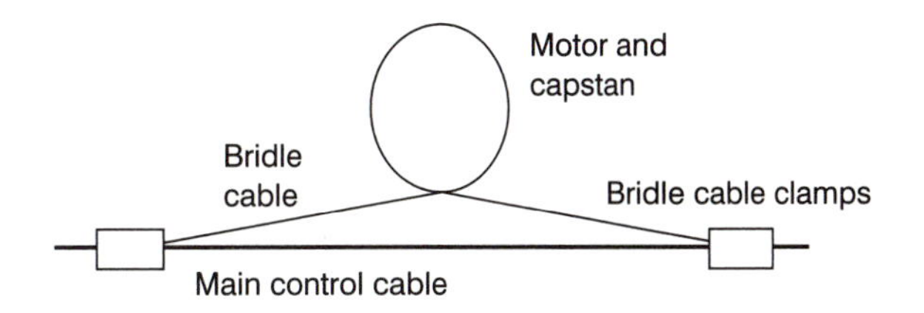

BASIC AUTOPILOT CONCEPTS

A single rule applies to almost all operations of an autopilot system—the rule of balancing of opposite errors. If this concept is understood, then most of the modes and operations of an AFCS (Automatic Flight Control System) are clear. For example, in the case of a wing leveler single-axis autopilot system, two signal voltages are compared, and if the difference between these signals is 0, no drive voltage is sent to the control surface drive motor (servo motor). These two voltage signals are the command signal (in this case a reference of 0 V indicating level flight) and the position signal. Usually the position signal is measuring where the aircraft is positioned in the roll axis.

Wing Leveler Autopilot

In the wing leveler autopilot, the command signal and the position signals are compared and if a deflection of the aircraft attitude is sensed, this creates the error signal. The error signal can be either positive or negative depending on the direction that the aircraft has rolled. The error signal is amplified in the servo amplifier and causes the servo motor to deflect the control surface through a capstan or other mechanical linkage.

This action, in and of itself, is not enough to provide smooth action of the autopilot system. The following is an example of the problem. For all of the 10 following steps, the command signal is a reference of 0 V, indicating that level flight is desired.

1. The aircraft is sitting in level flight. The roll position signal is 0 V. The servo drive is also 0 V so the servo motor is not turning. The ailerons are centered.

2. A wind gust deflects the roll axis of the aircraft to a position of a right bank of 10°. The roll position signal is +10. The error signal is +10. The servo drive is positive and the servo motor starts to drive. The

aileron position starts to deflect in the direction to return the aircraft to level flight (left roll). The aircraft starts to roll back toward level.

3. The aircraft is in a 9° right bank. The roll position is +9. The servo drive is positive and the servo motor is driving. The aileron position continues to deflect in the direction to return the aircraft to level flight. The returning roll action accelerates.

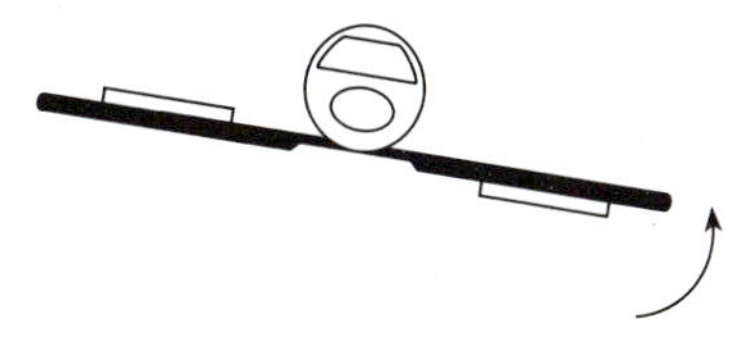

4. The aircraft is in a 5° right bank. The roll position signal is +5. The servo drive is still positive and the servo motor is still driving. The aileron position continues to deflect in the direction to return the aircraft to level flight. The returning roll action accelerates more.

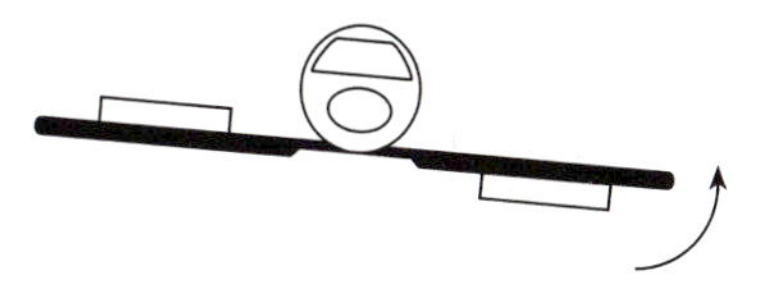

5. The aircraft is in a 1° right bank. The roll position signal is +1. The servo drive is still positive and the servo motor is still driving. The aileron position continues to deflect in the direction to return the aircraft to level flight. The returning roll action accelerates more.

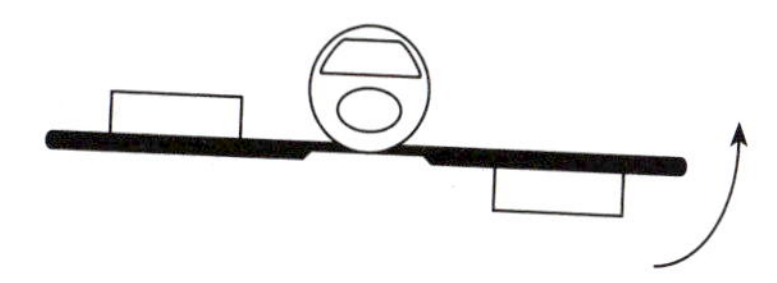

6. The aircraft is level. The roll position signal is 0. The servo drive is 0 and the servo motor is not driving. The aileron position is still deflected in the direction to return the aircraft to level flight (left roll). The left roll

action is still active. This causes the aircraft to overshoot the level position of the roll axis and continue rolling to the left.

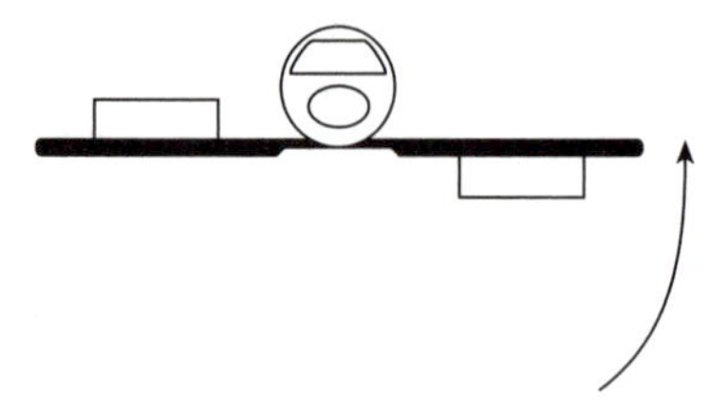

7. The aircraft is in a 1° left bank. The roll position signal is –1. The servo drive is negative and the servo motor is starting to drive in the other direction. The aileron position is still deflected in the direction to produce a left roll, but at a lower deflection. The left roll action slows down.

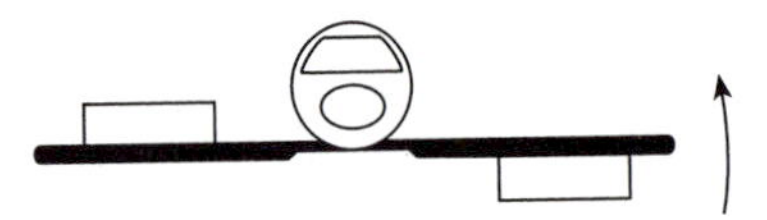

8. The aircraft is in a 5° left bank. The roll position signal is –5. The servo drive is negative and the servo motor is still driving in the other direction. The aileron position is still deflected in the direction to produce a left roll, but at a lower deflection. The left roll action slows more.

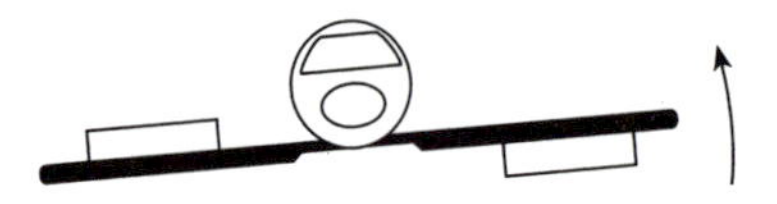

9. The aircraft is in a 10° left bank. The roll position signal is –10. The servo drive is negative and the servo motor is still driving in the other direction. The aileron position is centered. The left roll action stops.

10. The aircraft is in a 10° left bank. The roll position signal is –10. The servo drive is negative and the servo motor is still driving in the other

direction. The aileron position starts to move toward a right roll. The left roll action starts to the right.

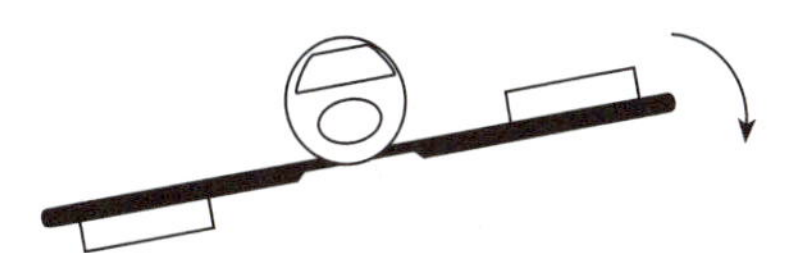

Position Control Feedback

As you can see, this will result in an oscillation back and forth in the roll axis. This problem is overcome with the use of either rate feedback or position control feedback. In rate feedback, a rate signal representing the speed of the servo motor is summed with the position signal and the command signal in such a way as to cancel the error signal. This allows for very high gain of the system when the motor is turning slowly, but prevents overshooting of the system when the motor is turning quickly. Position feedback is a signal that sums with the position signal and the command signal. The following seven sketches give an example of the previous sequence, but with position control feedback.

1. The aircraft is sitting in level flight. The roll position signal is 0 and the servo drive is 0 so the servo motor is not turning. The ailerons are centered.

2. A wind gust deflects the roll axis of the aircraft to a position of a right bank of 10°. The roll position signal is +10. The error signal is +10. The servo drive is positive and the servo motor starts to drive. The aileron position starts to deflect in the direction to return the aircraft to level flight. The aircraft starts to roll back toward level.

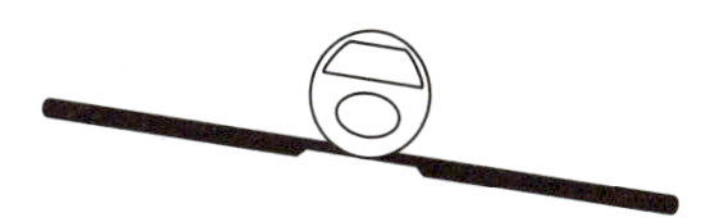

3. The aircraft is in a 9° right bank. The roll position signal is +9. The servo drive is positive and the servo motor is driving. The aileron posi-

tion continues to deflect in the direction to return the aircraft to level flight. The position control feedback signal is –4, causing the error signal to be +5. The returning roll action accelerates, but not as quickly as without position control feedback.

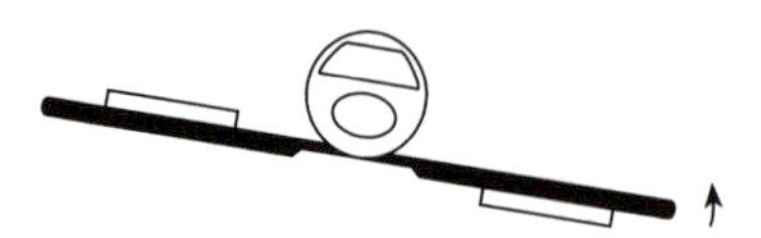

4. The aircraft is in a 5° right bank. The roll position signal is +5. The position control feedback signal is –5, causing the error signal to be 0. The servo drive has stopped and the servo motor is not driving. The aileron position is still deflected in the direction to return the aircraft to level flight. The returning roll action stabilizes.

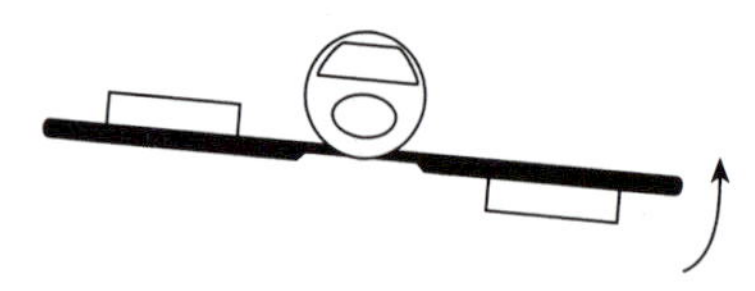

5. The aircraft is in a 2° right bank. The roll position signal is +2 volts. The position control feedback signal is –5, causing the error signal to be –3. The servo drive has reversed and the servo motor is driving in the opposite direction. The aileron position is moving toward neutral. The returning roll action slows down.

6. The aircraft is in a 1° right bank. The roll position signal is +1. The position control feedback signal is –2, causing the error signal to be –1. The servo drive is still reversed and the servo motor is still driving in toward neutral. The aileron position continues to move toward neutral. The returning roll action slows down more.

7. The aircraft is sitting in level flight. The roll position signal is 0 and the servo drive is 0 so the servo motor is not turning. The ailerons are centered.

This series of sketches shows the basic stability for level flight but does not allow for pilot input to control the autopilot for turn commands or other lateral modes. Getting back to the concept of balancing of opposite errors, it can be seen that any command signal that is summed with the position signal and position control feedback signal or rate signal will allow for several other modes of operation. Several examples of lateral modes of AFCS operation are Turn command, Heading mode (HDG), NAV mode, and Approach mode.

Turn Command and Other Modes

The *turn* command input is a good way to illustrate the balancing of opposite errors concept. The following five sketches show a sequence of steps for a pilot-selected turn using the turn knob input:

1. The aircraft is sitting in level flight. The turn command knob is centered so the turn command signal is 0. The roll position signal is 0 and the servo drive is 0 so the servo motor is not turning. The ailerons are centered.

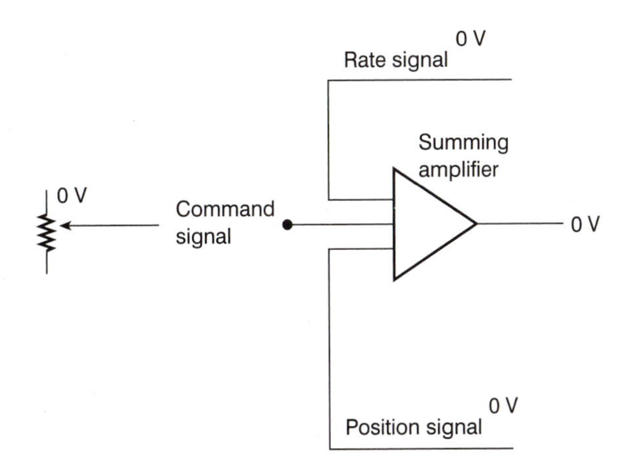

2. The pilot selects a right turn by rotating the turn control knob. This creates a turn command voltage of –10. This signal is summed with the rate signal (0) and the position signal. This results in an error signal of –10 and the servo motor starts to drive toward a right turn.

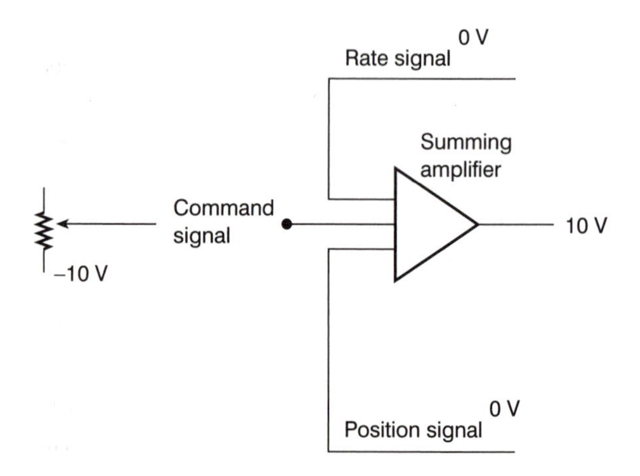

3. The aircraft rolls to 1° right. The command voltage is still –10, the position voltage is +1, the rate signal is +4. This results in an error of –5. The servo motor still drives toward a right turn, but not as quickly as before. The right roll accelerates.

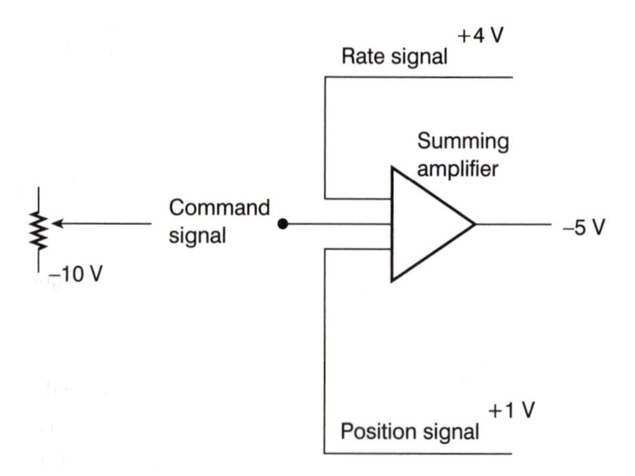

4. The aircraft rolls to 5° right. The command voltage is still –10, the roll position voltage is +5, the rate signal is +5. This results in an error sig-

nal of 0. The servo motor stops turning and the aircraft continues to roll to the right.

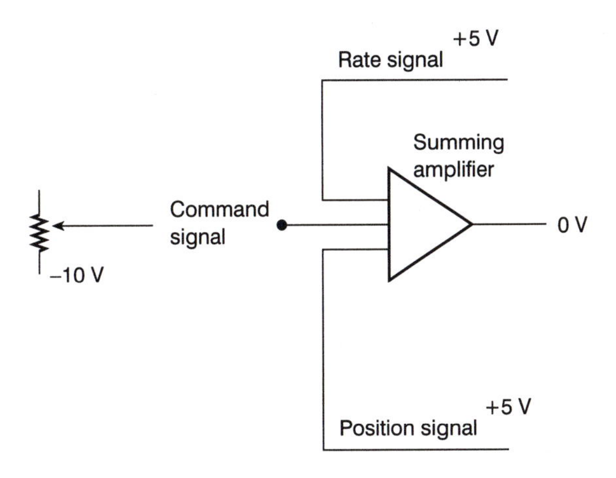

5. The aircraft rolls to 10° right. The command voltage is still –10, the roll position voltage is +10, the position control feedback signal is 0. This results in an error signal of 0. The servo motor is not turning, and the aircraft is now stable with a 10° roll to the right.

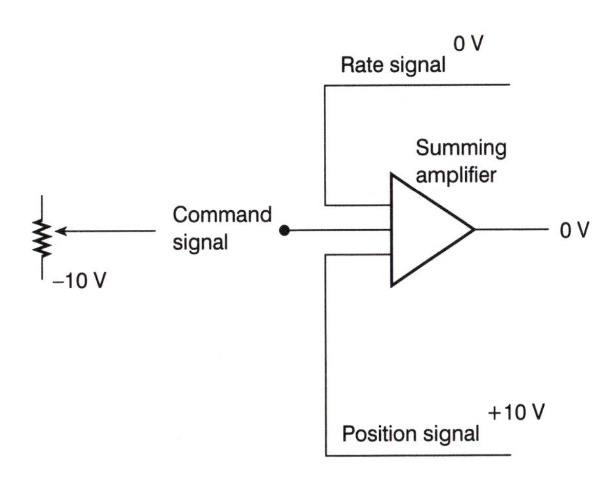

Another way of thinking of the balancing of opposite errors concept is to picture the aircraft in a stable 10° bank to the right, and a command of 10° right bank as was just shown. The voltage that is from the turn command knob is of a certain value, telling the autopilot to turn right. The vertical gyro is producing a voltage telling the autopilot to turn left in order to return to a level condition. If these voltages are exactly equal, they will cancel each other out and the result is an error voltage of 0. This way of thinking as it relates to the opposite errors concept can be expanded to the other lateral modes as well.

Heading mode is a mode of operation by which the pilot can select a desired compass heading. The AFCS will compare the aircraft's compass reading to the selected heading, and if they do not agree, will produce a heading error signal.

Heading mode has an input from the heading selector and an input from the Compass system. If the aircraft is headed east and the pilot selects a heading of north, a voltage is created that results in a left bank turn of some maximum value. Assuming this value is 25°, the aircraft will roll left until a bank angle of 25° is reached. Once again there is a balancing of errors, that is, a roll right signal from the vertical gyro and a roll left signal from the heading error.

A stable left turn is maintained until the aircraft heading gets close to north. When this happens, the heading error voltage reduces, causing the roll right command from the vertical gyro to become stronger by comparison and the aircraft rolls right, thus decreasing the amount of left bank, and the turn to the left slows down. If the autopilot is correctly adjusted, the bank angle reduces to 0 just as the aircraft rolls out on a heading of north.

NAV mode is similar except that an additional signal is summed in the process. In the Figure 16.3 example, the aircraft selects a VOR radial of 090, and the aircraft is located southeast of the VOR station on a heading of 090.

The NAV mode is engaged. The inputs to the summing point are 0 from the vertical gyro, turn left from the VOR deviation bar, and 0 from the course datum error.

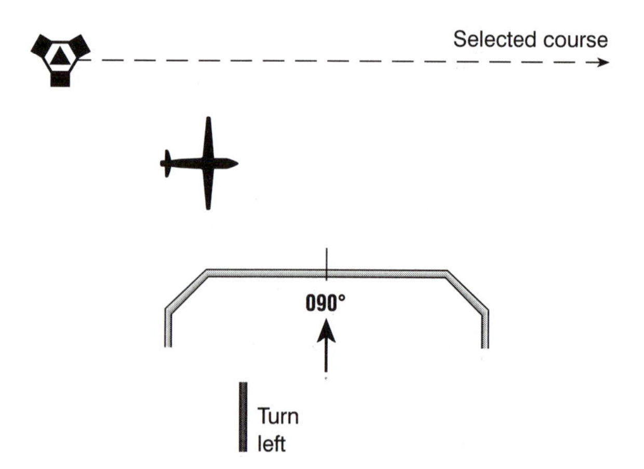

Figure 16.3
Autopilot—NAV Mode. Aircraft is right of selected course.

The *course datum error* is similar to the heading error in that it is a comparison of the compass heading and the selected course information from the navigation indicator.

This results in the aircraft establishing a left turn. As this turn is maintained, the compass heading decreases. This causes the course datum error to build in such a way as to call for a turn to the right.

The aircraft levels out when the "turn left" command from the deviation bar is equal to the "turn right" command of the course datum error. The aircraft maintains this heading until it gets close to the selected VOR radial, as shown in Figure 16.4. At this time, the deviation bar starts to center, and the amount of "turn left" command voltage reduces. This makes the "turn right" command stronger by comparison, and the aircraft rolls to the right. When the autopilot is correctly adjusted, this results in the aircraft rolling back to level, just as the aircraft arrives on the selected radial and heading (see Figure 16.5).

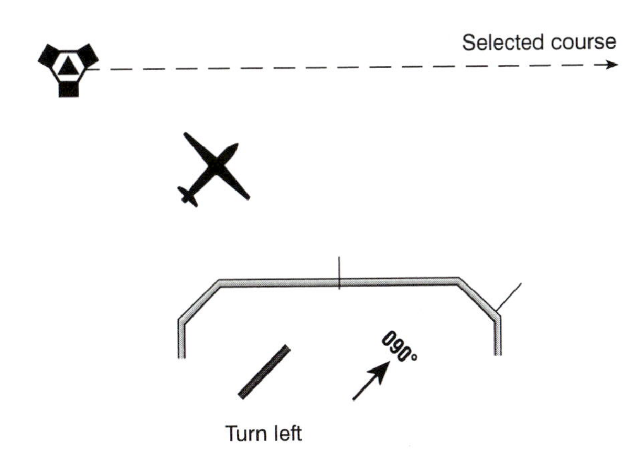

Figure 16.4
Autopilot—NAV Mode. Aircraft turning toward selected course.

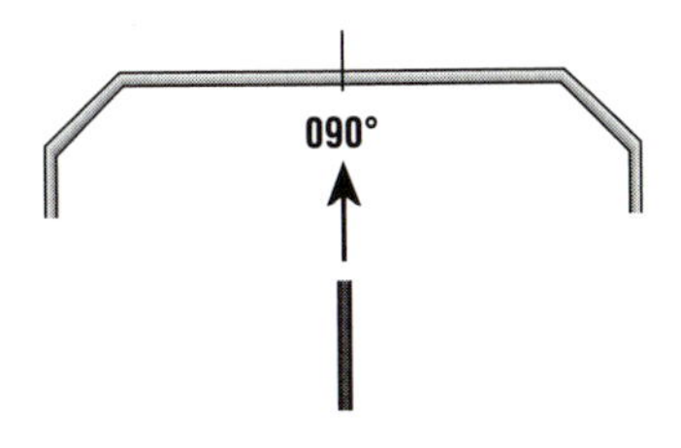

Figure 16.5
Autopilot—NAV Mode. Aircraft on selected course.

If the gain of the system or any of the inputs to the summing point are incorrectly adjusted, this will result in the aircraft either not capturing the desired heading or radial or oscillating around the correct heading or radial by overshooting the correct position and hunting back and forth.

QUESTIONS

1. What is a servo loop?
2. What is the difference between an autopilot and an AFCS?
3. What is the purpose of the pitch trim portion of an autopilot?
4. What is the fundamental difference between the roll channel reference and the pitch channel reference?
5. What would be the result if an autopilot were to malfunction such that the rate feedback signal were to be a constant 0 V?

APPENDIX A

Radio Frequency Spectrum

From 3 kHz to 30 GHz

RF Band	Range		Use
VLF	3 kHz	30 kHz	Navigation (NAV)
LF	30 kHz	300 kHz	NAV, broadcast, maritime
MF	300 kHz	3000 kHz	Broadcast, maritime
HF	3 MHz	30 MHz	CB, broadcast, maritime
VHF	30 MHz	300 MHz	FM, TV, NAV
UHF	300 MHz	3000 MHz	TV, radar, NAV, space, meteorological
SHF	3 GHz	30 GHz	Space, satellite, NAV, radar

APPENDIX B

Decibels and Power Relationship in Radar Transmission, Etc.

Consider the radar transmitter as sending out a 1000 W burst of RF energy. We can establish a reference level by using the equation:

$$\text{decibels} = 10 \log \frac{P_{out}}{P_{in}}$$

$$\text{dB} = 10 \log \frac{1000\ \text{W}}{1000\ \text{W}}$$

$$\text{dB} = 10 \log 1$$

$$\text{dB} = 0$$

In other words, the 1000 W is now established as a 0 dB reference level to which other signals can be compared.

Consider the energy being returned to the antenna by the aircraft. Using the same equation, the relationship between dB and power is this:

$$\text{dB} = 10 \log \frac{P_{out}}{P_{in}}$$

and modifying it to relate to the returned and the transmitted powers to compare the transmitted energy with the skin return energy

$$\text{dB} = 10 \log \frac{P_{out}}{P_{in}}$$

$$\text{dB} = 10 \log \frac{0.25\ \text{W}}{1000\ \text{W}}$$

$$\text{dB} = 10 \log 0.00025$$

$$\text{dB} = -36$$

The returned signal is then said to be –36 dB. This relieves us from concerning ourselves with actual wattages and allows us to compare signals with a common reference. We could now say that a transponder returns a signal that is +20 dB as compared to a normal skin return from radar of –36 dB. This means that the signal received at the ground radar antenna is 56 dB larger when from a transponder than when from radar.

Now using the equation in reverse:

$$\text{dB} = 10 \log \frac{P_{\text{out}}}{P_{\text{in}}}$$

$$56 \text{ dB} = 10 \log \frac{Pr}{P2}$$

$$56 \text{ dB} = 10 \log \frac{Pr}{1000 \text{ (reference wattage)}}$$

$$\frac{56}{10} = \log \frac{Pr}{1000}$$

$$5.6 = \log \frac{Pr}{1000}$$

$$398107 = \frac{Pr}{1000}$$

$$398107171 = Pr$$

The signal returned from the transponder is 400 million times larger than from normal skin return radar!

Example: If the power input is 100 W and the power output from the system is 50 W, then the ratio

$$\frac{P_{\text{out}}}{P_{\text{in}}} = 0.5$$

$$\text{attenuation} = -3 \text{ dB}$$

This indicates that the power has decreased by half, or the signal has been attenuated from a normalized level down 3 dB. When the transponder returns a signal that is 20 dB greater than a normal skin return this means:

$$20\text{ dB} = 10\log\frac{P_{\text{out}}}{P_{\text{in}}}$$

$$\frac{20}{10} = \log\frac{P_{\text{out}}}{P_{\text{in}}}$$

$$2 = \log\frac{P_{\text{out}}}{P_{\text{in}}}$$

$$100 = \frac{P_{\text{out}}}{P_{\text{in}}}$$

Therefore the power returned by the transponder is 100 times larger than a normal skin return as would be the case with a normal radar echo.

APPENDIX C

Binary to Octal Conversions

The short conversion method:

Group the Binary bits in groups of three.
Work left from the decimal point.
Convert each group of three bits into a decimal equivalent.
The result is the Octal number

The select switch allows the pilot to select up to a number 7777 on the transponder. This is actually an Octal number.

Let's convert this Octal number to Binary.

7	7	7	7	Octal
111	111	111	111	Binary

When we convert this Binary number to decimal format we have

111111111111.0

Converting this requires adding the weighting of each position, so we add

$$1 + 2 + 4 + 8 + 16 + 32 + 64 + 128 + 256 + 512 + 1024 + 2048$$

This gives us the total possible number of combinations in decimal 4096. (Don't forget to add the 0 condition.)

Altitude Reporting Code Chart for Transponder Mode *C*

Range					Pulse Position						
*X*1000	*D*2	*D*4	*A*1	*A*2	*A*4	*B*1	*B*2	*B*4	*C*1	*C*2	*C*4
26.3	0	0	1	0	1	1	0	0	1	0	0
26.4	0	0	1	0	1	1	0	0	1	1	0
26.5	0	0	1	0	1	1	0	0	0	1	0
26.6	0	0	1	0	1	1	0	0	0	1	1
26.7	0	0	1	0	1	1	0	0	0	0	1
26.8	0	0	1	0	0	1	0	0	0	0	1
26.9	0	0	1	0	0	1	0	0	0	1	1
27.0	0	0	1	0	0	1	0	0	0	1	0
27.1	0	0	1	0	0	1	0	0	1	1	0
27.2	0	0	1	0	0	1	0	0	1	0	0
27.3	0	0	1	0	0	1	0	1	1	0	0
27.4	0	0	1	0	0	1	0	1	1	1	0
27.5	0	0	1	0	0	1	0	1	0	1	0
27.6	0	0	1	0	0	1	0	1	0	1	1
27.7	0	0	1	0	0	1	0	1	0	0	1
27.8	0	0	1	0	0	1	1	1	0	0	1
27.9	0	0	1	0	0	1	1	1	0	1	1
28.0	0	0	1	0	0	1	1	1	0	1	0
28.1	0	0	1	0	0	1	1	1	1	1	0
28.2	0	0	1	0	0	1	1	1	1	0	0
28.3	0	0	1	0	0	1	1	0	1	0	0
28.4	0	0	1	0	0	1	1	0	1	1	0
28.5	0	0	1	0	0	1	1	0	0	1	0
28.6	0	0	1	0	0	1	1	0	0	1	1
28.7	0	0	1	0	0	1	1	0	0	1	1
28.8	0	0	1	0	0	0	1	0	0	0	1
28.9	0	0	1	0	0	0	1	0	0	1	1
29.0	0	0	1	0	0	0	1	0	0	1	0
29.1	0	0	1	0	0	0	1	0	1	1	0
29.2	0	0	1	0	0	0	1	0	1	0	0

APPENDIX E

Gray Code and Conversions

The Gray code belongs to a class of codes called minimum change codes.

Example: only one bit changes when incrementing or decrementing by one step.

Unweighted code bit positions in the code do not have any specific weight assigned to them, and therefore cannot be used for arithmetic operations.

Decimal	**Binary**	**Gray**
0	0000	0000
1	0001	0001
2	0010	0011
3	0011	0010
4	0100	0110
5	0101	0111
6	0110	0101
7	0111	0100
8	1000	1100
9	1001	1101
10	1010	1111

Application—analog to digital and digital to analog conversion; three-bit code wheel converts analog to digital.

Analog-to-Digital Code Wheel

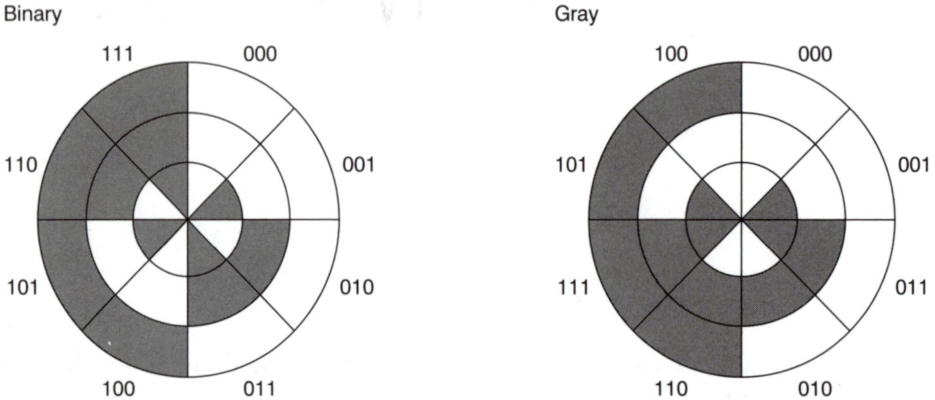

Binary

111 000 001 010 011 100 101 110

Gray

100 000 001 011 010 110 111 101

Explain how a misread bit in Binary could result in a max of 180° error.

Example: When going from 011 to 100, if 1st MSB (Most Significant Bit) goes high before the 2nd and 3rd MSB goes low, a 180° error occurs.

Conversion: Binary to Gray

Rules:

1. MSB(Binary) = MSB Gray
2. 1st MSB + 2nd MSB = 2nd Gray bit
3. 2nd MSB + 3rd MSB = 3rd Gray bit

This is called Modulo 2 addition. It is Binary addition with the carry ignored and is accomplished with an "exclusive or" logic function.

Example:

Convert

$101_{(2)}$ to Gray
1 1 1 compare to the code wheel for proof

Sample problems:

$1\ 0\ 0\ 1\ 1\ 0\ 0\ 1_{(2)}$ to Gray
1 1 0 1 0 1 0 1

$1\ 0\ 0\ 1\ 1\ 1\ 1\ 1_{(2)}$ to 1 1 0 1 0 0 0 0 Gray 159
$1\ 0\ 1\ 0\ 0\ 0\ 0\ 0_{(2)}$ to 1 1 1 1 0 0 0 0 Gray 160

Note: Only one bit is changed in Gray code as it is incremented. Binary code changes six bits for one increment.

Gray to Binary

Rules:

1. MSB (Gray) = MSB (Binary)
2. 1st Binary + 2nd Gray = 2nd MSB (Binary)
3. 2nd Binary + 3rd Gray = 3rd MSB Gray

This is called Modulo 2 addition. It is Binary addition with the carry ignored and is accomplished with an "exclusive or" logic function.

Example:

Convert

1 0 1 0 Gray to Binary
$1\ 1\ 0\ 0_{(2)}$

Convert

1 1 0 1 0 0 0 0 Gray
$1\ 0\ 0\ 1\ 1\ 1\ 1\ 1_{(2)}$ 159

APPENDIX G

VLF Stations Available for Omega Supplemental Usage

Call Letters	Location	Frequency, kHz
NAA	Maine	17.8
NSS	Maryland	21.4
NBA	Panama	18.6
NLK	Washington	23.4
NPM	Hawaii	23.4
NWC	Australia	22.3
GBR	England	16.0
NDT	Japan	17.4
JXM	Norway	16.4

SOLUTIONS

Chapter 1: ILS

1. d
2. a
3. b
4. c
5. d
6. b
7. c
8. a
9. b
10. c
11. b
12. a
13. d
14. d
15. d
16. Localizer = 112.3
    Marker beacon = 75 MHz
    Glide slope = 334.7 MHz
17. See Figures 1.1, 1.2, and 1.4.

Chapter 2: MLS

1. 0 – ± 40°; can be expanded to ± 60°
2. 0.9 – 7.5°; can be expanded to ± 20°
3. 200
4. 5031–5095 MHz
5. Allows flexible approach path; more available channels; increased safety standards.

6. DME
   Approach azimuth
   Approach elevation
   Enhancement
7. See Figure 2.9.
8. Antenna locations around the runaway:

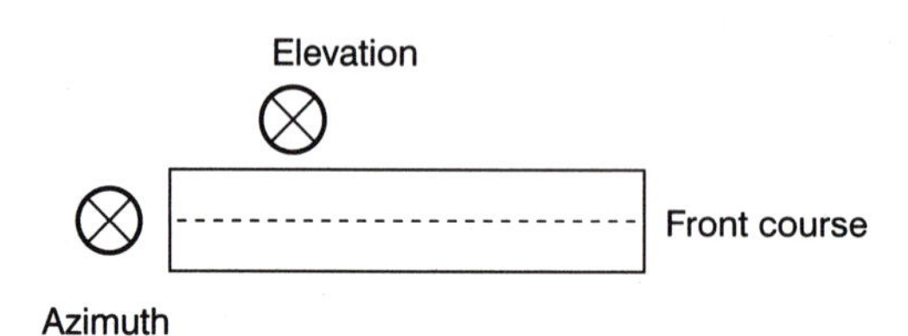

9. One forward on nose
   One aft on belly

Chapter 3: RMI

1. c
2. d
3. d
4. d
5. f
6. g
7. e

Chapter 4: HSI

1. d
2. a
3. e
4. d
5. c
6. a
7. b
8. c
9. a
10. e

Chapter 5: VOR

1. a
2. a
3. b
4. b
5. True
6. True
7. a

Chapter 6: ADF

1. False
2. False
3. True
4. True
5. False
6. True
7. True
8. False
9. True
10. True

Chapter 7: DME

1. a
2. b
3. b
4. 12.36 μs per NM round trip
5. 12 μs
6. 12 μs
7. 36 μs
8. 30 μs
9. Yes, a Morse code, on occasion at 1350 Hz.
10. Squitter is the random pulses received by the DME receiver.
11. Slant range error is greatest when the aircraft is directly over the DME at high altitudes.

12. Slant range error is least when the aircraft is at low altitude and long range from DME.
13. 2700
14. In search mode there is a high pulse repetition frequency of 115 PRF looking for the correct reply. Warning flag shows that no reliable DME range is available for display. In track mode there is a low pulse repetition frequency of 25 PRF indicating DME has found the correct reply, DME range is displayed, and the DME flag is retracted.
15. 50 μs
16. DME "hold" function allows the DME to retain the information about the last frequency that it was dialed to, even when the NAV has been rechannelled.
17. The suppression bus prevents interaction between the DME system and the transponder system.
18. As it is maneuvered, the aircraft itself can shield the line of sight transmission and cause loss of signal. In the airborne circuitry a memory holds the last calculated data for about 20 sec before the loss of signal is reported to the display indicator. If the signal is lost then the DME coasts for 10 sec before the circuitry decides that it has lost tracking, then initiates the search and lock-on procedures
19. DME is channelled by the VHF navigation head automatically.

Chapter 8: RNAV

1. RNAV is an electronic method of placing pseudo waypoints into a flight plan and using them as though they were real VORTACs and/or VOR/DMEs.
2. A fix can be established using either technique—the Rho Theta, the Rho Rho, or a combination of both. Set OBS course selector to the bearing. Fly the needle centered. DME displays distance and time to station.

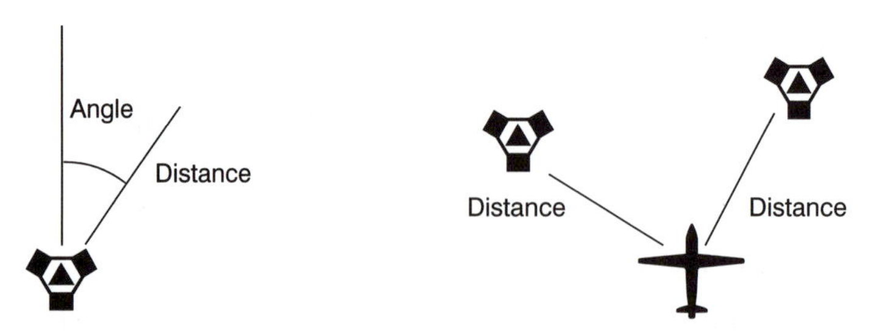

3. Shorten the actual enroute distance and therefore time.
   Affect fuel costs.

Reduce traffic at VORTACs and other centers.

Provides a constant VOR course width.

4. An electronically generated VOR/DME station that appears real to the pilot.
5. Yes . . . it can be positioned anywhere as set by the flight crew to the computer.

   No . . . only within the electronic limits of the VOR/DME.

   The Waypoint will function as a Waypoint only if real VOR/DME signals are present.

   If the Waypoint is positioned outside of the actual reception area of the real VORTAC, then the CDI D bar will not operate and a fault flag will occur, even though the pilot has selected the Waypoint outside of the reception area.
6. No . . . RNAV microprocessor defaults to VOR/DME mode.
7. Aid to enroute navigation—runway location for non-ILS approach
8.

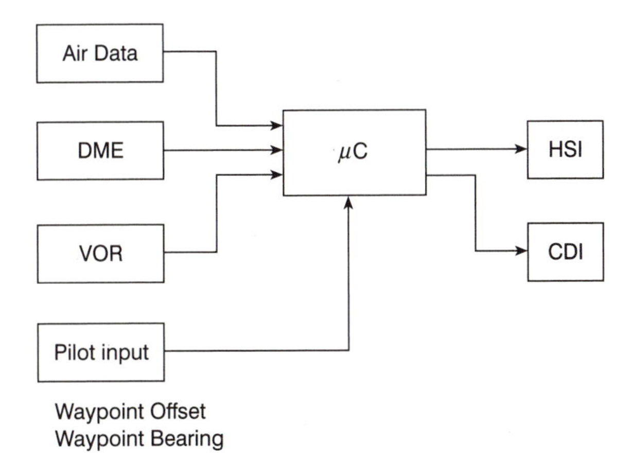

9. In the RNAV mode, the CDI displays DISTANCE information to the selected course. As the aircraft approaches the Waypoint the CDI needle will close. The full-scale deflection of the CDI will be about 5 NM and in the approach mode the CDI will display full-scale deflection of 1.25 NM.
10. VOR—bearing information

    DME—distance information

    Barometric pressure—altitude information
11. Yes

Chapter 9: WX

1. High-frequency RF wave bouncing back from a target
2. 3 to 300 GHz
3. 12.36 μs
4. Cathode ray tube
5. Color patterns, with magenta usually of the highest severity
6. Allows pilot control of the pitch axis of the radar antenna
7. 2 to 250 NM

Chapter 10: ATCTX

1. 1030 MHz
2. 1090 MHz
3. An interrogation from a side lobe transmission instead of the main beam
4. This is the interrogation from the SSR. This pulse has the most power.
5. By referencing the amplitude of the *P*1 and *P*2 pulses
6. Mode *C*
7. A pilot-selected code and altitude
8. None
9. Transponder code sequence for 26,500 ft of altitude:

	*D*2	*D*4	*A*1	*A*2	*A*4	*B*1	*B*2	*B*4	*C*1	*C*2	*C*4
26,500 ft	0	0	1	0	1	1	0	0	0	1	0

10. To ensure that there can be a positional error of only 1 bit instead of a possible 9-bit error (about 200 ft error)
11. Radio Detection and Ranging
12. To provide a positive identification to the ATC. The VHF comm is used to command a squawk ident. The target is then radar identified.
13. An electronic request for an altitude code; Mode *A,* Mode *C,* etc.
14. Spoken by the air traffic controller, who asks the pilot to identify his aircraft; pilot presses ident button, radarscope blossoms at aircraft position

15. Called framing pulses; identify the beginning and the end of the transmit sequence
16. Belly, centerline, as far away from the DME antenna as is possible
17. 4096_{10} 7777_{8}
18. Decibels
19. To avoid interaction with the DME (same frequency band); either ATCTX or DME transmits at a given time
20. For the expeditious movement of aircraft
21. Area of ATC—all aircraft in zone must report to ATC and receive an ATC clearance regarding proposed movement; about 5 to 10 NM, altitudes vary ≈ 10,000 ft
22. Reply light flashes when transponder is replying to ground interrogation.
23. An authorization from ATC to move aircraft in proposed direction and altitude
24. Pulse code for onboard emergency (7700):

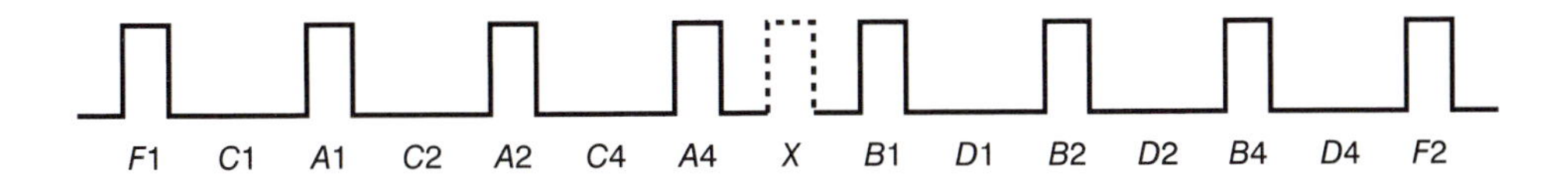

25. Pulse code for a selected code of 2563:

	A	*B*	*C*	*D*
Octal	2	5	6	3
Octal coded	010	101	110	011

Converted to pulse sequence:

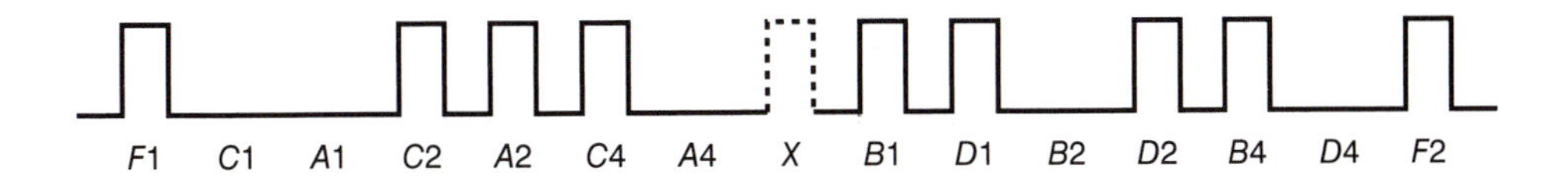

26. Select

Chapter 11: TCAS

1. True
2. True
3. True
4. True
5. False
6. True
7. Joint Enroute Terminal System (Canadian)
   Terminal Radar Approach and Control (American)
8. Old equipment, overloaded operations; current system is reactionary to existing conditions rather than predictive; human operators
9. In USA: situation has become desperate and critical; in Canada: situation is lagging behind but is still a problem in large centers
10. The 10-year plan is to move from human controllers to a computerized system. This will change the principle of the system from one of reaction to one of prediction. Current situation is to react almost immediately to crisis situations as they occur. The AERA system will be able to predict conflicts up to 20 min ahead of time, and solve the rerouting problem before it actually occurs.

Chapter 12: Loran C

1. 100 kHz
2. $5990 \times 10 = \text{GRI in } \mu s$.
3. To ensure that the master station is always received first.
4. Long Range Navigation
5. Three stations—two slave and one master
6. 22
7. A triad—one master and two slaves
8. No. There are no Loran receivers that are certified for this.
9. Group Repetition Interval
10. Digital display of latitude and longitude
11. Generally ± 100 ft
12. Ground wave
13. V, W, X, Y, Z

14. Skin mapping is a surface analysis of the aircraft's magnetic fields to find where antennas interfere with each other, or where there are unusually large areas of magnetic influence that interfere with normal transmit/receiver operation.
15. By its GRI

Chapter 13: Omega

1. d
2. b
3. a
4. a
5. d

Chapter 14: ARINC 429 Bus

1. Discrete, numeric, alphabetic, graphic
2. Letters, abbreviations, punctuation
3. 12.0 kHz and 14.5 kHz
4. +10 V
5. –10 V
6. Zero
7. Add all of the bits in a word to create a word that has an odd number of bits; used to ensure integrity of the word transmission.
8. Shielded twisted pair
9. Digital Information Transfer System

Chapter 15: GPS

1. GPS was designed to provide an accurate worldwide three-dimensional positioning system.

   Transmission received from satellites enable precise timing of radio transmission from satellites to ground stations. The times and therefore distances of three satellites' positions allow accurate three-dimensional telemetry and therefore accurate navigation fix in latitude, longitude, and altitude.

2. The space segment: orbiting satellites

   The control segment: master control with uplink capability

   The user segment: one-, two-, and multi-channel capability
3. 21 plus three spares for a total of 24
4. 5 stations with 1 master and 1 backup master
5. Colorado Springs (master); Onizuka Air Force, California (backup master); Ascension Island (South Atlantic), Diego Garcia Island (Indian Ocean), Kwajalein (South Pacific), Hawaii, and Cape Canaveral (Florida).

   Provide uplink and satellite trajectory correction ability.
6. One-channel receivers receive data from 4 satellites but in a sequential order; two-channel receivers receive data from 4; but simultaneously to two channels and multi-channel receivers accept data from 4 (even to 6) satellites simultaneously.
7. Approximately 12 hr; 10,900 NM.
8. Four
9. L1: 1575.42 MHz

   L2: 1227.6 MHz
10. The accuracy of the system depends on operational mode.

    Generally it is 1 m for military applications and as bad as 100 m for civilian application.

    It is also possible in geological survey applications to have accuracies as good as ± 1 cm.
11. The Russian equivalent to the GPS.

    Both sets of satellites will be used for combined navigation purposes once some technical details about geometric grids and latitude and longitude formats have been approved by FAA.
12. 24 hr per day.
13. Five—four for navigation tracking and one for inputting ephemeris and almanac data.
14. Three
15. One
16. Receivers rely on the integrity and fidelity of the satellites constellation; military has the option to degrade the system.
17. Approximately 12.5 min
18. Approximately 30 sec

Chapter 16: Autopilot

1. A servo loop is a system by which the position of a load is controlled by a motor drive that is derived by comparing a signal representing load position to the desired position.
2. The difference between an autopilot and an AFCS is that an AFCS includes a flight director as well as an autopilot.
3. The purpose of the pitch trim portion of an autopilot is to decrease the long-term loads on the pitch servo due to the changing trim of the aircraft.
4. The roll channel uses 0° as its reference at all times because any other angle will cause a left or right turn. The pitch channel must be constantly changing depending on center of gravity, flap position, gear up or down, etc.
5. The aircraft would oscillate about the axis controlled by the servo motor that has lost the rate feedback signal.

GLOSSARY

Above Ground Level (AGL). Height above the ground, always understood to mean ft.

A/C. Aircraft.

acquisition time. Time required for a receiver to lock on to a given signal.

ADC. Air Data Computer.

ADF. Automatic Direction Finder Navigation System.

ADI. Attitude director indicator.

AERA. Automated Enroute Air Traffic Control.

AFCS. Automatic Flight Control System.

AFDS. Automatic Flight Director System.

AGC. Automatic Gain Control. An electronic feedback process.

AGL. *See* **Above Ground Level.**

Air Data Computer. Computer that handles all air data information for altitude processing and any other system on board that requires air information.

airways marker. 3000 Hz, white marker panel light, Morse code dot dot dot, possibly other Morse code identifier.

AM band. 550 kHz to 1650 kHz contains AM radio station frequencies.

amplitude modulation. The mixing of a carrier frequency (fc) and an audio frequency (faudio), usually. The amplitude variations of the carrier frequency vary in direct relation to the audio information. *See* **mixing.**

approach mode. The CDI has a full-scale calibration of 5 NM.

Area Navigation (RNAV). A guidance system that uses DME slant range, VOR bearing, and pilot inputs to compute bearing and distance to a Waypoint; the pilot flies to the Waypoint as though it were a real VORTAC station.

ARINC 429. A digital standard established by Aeronautical Radio Inc. in 1979; an aviation data bus.

ASIC. Application Specific Integrated Circuit.

ATC. Air Traffic Control.

ATC System. Allows for tracking of any aircraft, providing the Air Traffic Controller with aircraft type identification, registration, altitude, range, bearing, and airspeed.

ATC transponder (ATCTX). A transmitter/receiver unit that transmits in response to a ground interrogation; when requested to do so by Air Traffic Control, the pilot presses the "Ident" button and the ATC radar screen "blossoms," thus positively identifying any given A/C.

ATCRBS. Air Traffic Control Radar Beacon System.

attitude director indicator. *See* **ADI.**

automatic VOR. Compass heading and VOR radial information are combined to provide relative bearing information to be displayed on the RMI (or RMI pointer of the HSI).

autopilot. A system that employs position control; a part of the Automatic Flight Control system.

avionics. A contraction of the words *avi*ation electr*onics.*

azimuth. Horizontal parameter.

azimuth reference. Horizontal information.

back course. There exists a localizer radiation pattern from the back side of the active localizer system. This radiation pattern is similar to the front course with some notable differences. The glide slope information is *not* available on the back course approach. The 90 and 150 modulated tones are on the same side as front course and therefore the pilot must be aware that instead of the usual procedure of flying to the needle, in back course approach one must fly away from the needle. *See* **front course.**

barometric altitude. Altitude information received from barometric pressure instrument.

BCD. Binary Coded Decimal; coding a decimal number in a binary format.

bearing. This is a magnetic direction from the aircraft toward the station.

BFO. Beat frequency oscillator; used for interrupted carrier wave.

Bidirectional. RF wave propagation with signal strength strong in only two directions.

Binary. A number system based on two numbers.

Binary Coded Decimal. *See* **BCD.**

bipolar. In ARINC 429, the voltages are +10 = logic one, and –10 = zero, called bipolar.

BNR. Binary number system based on two numbers.

C band. Frequency range from 5031 MHz to 5095 MHz.

C VOR. Conventional VOR.

C/A code. Coarse Acquisition (coarse meaning approximate or rough).

cardioid. A resultant reception pattern of the sense and loop antennas.

cathode ray tube. *See* **CRT.**

CDI. Course Deviation Indicator; indicates VOR radial left or right of the selected bearing. The red flag indicates loss of a valid signal. The TO/FROM flag indicates the aircraft is either to or from the VOR station.

code selection 7 X X X. A "7" in the first number position of the four-digit Code Selection Number indicates an emergency situation.

compass card. Compass information and a pointer (fiducial) indicating the direction that the aircraft is pointed.

Compass Rose. A navigation aid shown on maps indicating compass points, usually in 1° increments.

Compass System. Provides primary "heading reference" (compass direction), acts as a source from flight control system, usually within ¾ of a degree.

Cone of Silence. Inverted cone-shaped area above a VOR ground station that results in erratic VOR operation.

Constellation. A collection of orbiting satellites.

Control Head. A panel device allowing pilot interaction with remote mounted units used to control mode, frequency, etc.

Control Segment. The ground controlling portion of the GPS.

Control Zone. An area surrounding a control tower (approximately 5 to 10 NM radius to 4000 AGL) where air traffic is controlled by the ATC facilities. All ground traffic arrival, departure, or flying through this zone is the responsibility of ATC.

Controlled Airspace. An airspace that has been defined and has air traffic control service available.

course. The direction that the aircraft is actually flying.

course selector. The portion of the HSI that performs the same task as the OBS (also autopilot output).

crabbing. *See* **heading.**

CRT. Cathode ray tube; an electron gun that directs high-speed electrons onto a phosphorescent screen.

D Bar. Deviation bar, track bar, or course deviation indicator.

D VOR. Doppler VOR.

dBW. Decibel with a 1 W reference signal.

dead reckoning. A method of navigation using time, speed, distance, and known starting position.

decimal. A number system based on 10 numbers.

Decision Height (DH). The altitude and visibility restrictions that determine the point at which the pilot must decide to either abort the landing or land the aircraft.

Dedicated Data Link. Data link from only one transmitter to only one receiver.

diplexer. An RF signal switch that directs signal flow; also called duplexer.

DITS. Digital Information Transfer System.

diurnal effect. The change in the ionosphere's altitude due to daytime/nighttime transition.

DME. Distance Measuring Equipment.

DME channelling. DME channel selection is done through the navigation receiver channel selector.

DME hold. Allows the DME to retain information about the last channel selected through the NAV head.

DME slant range. Actual direct line of sight distance from the interrogator (airborne) to the transponder (ground station).

Doppler effect. The frequency of the waves moving *toward* an object is higher than the actual propagating frequency. The frequency of the waves moving *away from* an object is lower than the actual propagating frequency.

duplexer. *See* **diplexer.**

E field. Electrostatic portion of a radio transmission. It is vertical in a vertically polarized transmission (perpendicular to earth's surface) and horizontal in a horizontally polarized transmission (parallel to earth's surface).

echo. The returned energy displayed on the CRT as an echo.

EFIS. Electronic Flight Instrumentation System.

elevation reference. Vertical information.

elevation. Vertical parameter.

enroute mode. The CDI has a full-scale calibration of 1.25 NM.

Envelope. With reference to TCAS, the envelope is the window of protected airspace—the space surrounding the aircraft that is considered to belong to the aircraft (about 3 mi in diameter and 750 ft both up and down from the plane). With reference to an aircraft, the envelope is the physical limits of the aircraft's performance.

FAA. US Federal Aviation Administration.

FAF. Final approach fix Waypoint.

fc. Carrier frequency.

FCS. Flight Control System (the combination of an autopilot system and a Flight Director System).

FDC. Flight Director Computer.

FDS. Flight Director System.

fiducial. Heading index, a fixed pointer.

fix. An aircraft's current location, sometimes called a "running fix" if it takes some amount of time to perform the calculation manually.

Flight Level. A method of identifying altitude; for example, "Flight Level 18" means 18,000 ft above ground level; for brevity the three zeros are dropped and the expression is "FL 18."

flight path. The ideal track of the aircraft both vertical and horizontal.

Framing pulses. F1 and F2 are beginning and end of reply transmission.

frequency agility. The ability of the electronics to select information from more than one transmitter and correlate these data into the "best case scenario." This best case information is then presented to the flight deck.

FRO scan. On azimuth, moves from the far left to the far right. On elevation, moves upward from the ground.

front course. Flying the localizer and glide slope to the runway is called a front course approach (*see* **back course**).

fruiting. When the ATC transponder is being interrogated by more than one ground station, replies to one ground station are received by the other.

garbling. Overlapping of the replies from two aircraft from the same interrogator. Also called **garble, sync garble,** or **synchronaes garble.**

General Data Link. Data link to many possible receiver locations.

glide path. The vertical track of an aircraft's movement.

glide slope (G/S). The ideal descent path on an ILS approach.

GLONASS. *Glo*bal *Na*vigation *S*atellite *S*ystem (Russian).

GMT. Greenwich Mean Time; the time of day at zero degrees longitude used as a world time reference.

goniometer. RF resolver. *See* **resolver.**

GPS. Global Positioning System (American).

Gray code. A number system with only a one-bit change from any number to the next; cannot be used for mathematical operations, only for data transmission.

Great Circle. The shortest distance between any two places on the surface of the earth.

GRI. Group Repetition Interval; example—West Coast GRI is 9940.

ground speed (GS). The speed over the ground.

ground transmission delay time. 50 microseconds (μs).

Ground Wave propagation. ADF frequency band 200 kHz to 1700 kHz.

GS. *See* **ground speed.**

G/S. *See* **Glide slope.**

H field. Electromagnetic portion of a radio transmission. It is perpendicular (transverse) to E field, horizontal in a vertically polarized transmission (parallel to earth's surface), and vertical in a horizontally polarized transmission (perpendicular to earth's surface).

Heading Bug. Often referred to as the bug; it is controlled from either the autopilot control panel or from the face of the HSI, and is used to set the desired course or as a reminder of the next course to steer.

heading (HDG). The direction in which the aircraft is headed. (Due to winds, it is possible to head in one direction and actually fly in another—**crabbing**).

HEX. Hexadecimal numbering system based on 16, using digits 0 to 9 and letters A, B, C, D, E, and F.

HF. High frequency; 3 MHz to 30 MHz.

HSI. Horizontal Situation Indicator; multipurpose navigation indicator that displays VOR, compass, and ILS information.

hyperbolic navigation. The process of locating a fix when information is received that is of a hyperbolic nature.

ideal flight path. The perfect horizontal and vertical path that will take the aircraft to the end of the runway.

IF. Intermediate frequency.

IFR. Instrument Flight Rules.

ILS. Instrument Landing System.

ILS channel limitation. Forty channels.

INS. Inertial Navigation System. An airborne self-contained global navigation system; requires no radio signals from stations external to the aircraft; relies on gyro gimbal assemblies and accelerometers for sensing position changes.

interrogation. The ATC ground station that transmits at 1030 MHz.

interrogator. An electronic system that sends out a code of information and then waits for a specific reply code.

IOC. Initial Operational Capability (in GPS).

ionosphere. Atmosphere surrounding the earth about 60 mi up.

IVSI. Modified Vertical Speed Indicator; sometimes referred to as instantaneous vertical situation indicator.

JETS. Joint Enroute Terminal System (Canada).

L Band. Frequency range from 978 MHz to 1215 MHz.

L Band (for GPS). L1 frequency is 1575.42 MHz; L2 frequency is 1227.6 MHz.

lambda $= 300 \times 10^6$ meters per second divided by the frequency; lambda is the wavelength of the frequency.

Lane. Width equal to the wavelength of the transmission frequency.

LF. Low frequency; 30 kHz to 300 kHz.

line of sight. Direct line of propagation of a signal.

localizer. The ideal azimuth path.

loop antenna. Ferrite core with two coils aligned at right angles.

Loran. *Lo*ng *Ra*nge *N*avigation; a long-range, low-frequency hyperbolic radio navigation system that uses time intervals instead of exact distances.

main beam interrogation. The main power pulse from the SSR.

MAP. Missed approach point Waypoint.

MF. Medium frequency; 300 kHz to 3000 kHz.

middle marker. Vertical signal cone, about 3500 feet from end of runway; pilot sees amber light, hears 1300 Hz audio tone Morse coded dot dash.

mixing. The formal term for combining two frequencies.

MLS. Microwave Landing System.

Mode *A*. Bearing and range information is transmitted.

Mode *B*. Allocated for European operation.

Mode *C*. Bearing, range, and altitude information is transmitted.

Mode *S*. Interrogation pulses are spaced differently than transponders spacing.

Morse code. Standard Morse code station identification signal.

multipath. One signal arriving at the receiver at two different times.

multipath error. There can be signal reflection from mountains, hills, etc. that cause distortion of the variable signal.

nautical mile (NM). Equal to 6076 ft, 1.15 statute mi, or 1852 m.

NAV. Navigation. **Nav** is also commonly used.

NAVAIDs. Ground stations specifically designed for ADF navigation.

NDBs. NonDirectional Radio Beacons; 200 kHz to 500 kHz; for use with the ADF system.

night effect. An ADF error caused by partial horizontal polarization of the incident waveform.

NORDO. Nonoperational radio situation.

NOTAM. Notice To Airman; an official notice published by Transport Canada providing advice to pilots.

null. An area or voltage that is at a minimum or zero value; often referred to as the null zone, and is used as the area of reference to tune the ADF receiver.

OBS. Omni Bearing Selector allowing pilot to select a bearing on the compass card.

OCI (Out of Coverage Indication) pulse. An MLS sector signal indicating that the aircraft is out of the scanning zone; can be left, right, or rear OCI.

Octal. A number system based on eight numbers.

Omega. A worldwide, low-frequency, long-range, hyperbolic navigation system.

Omni directional. RF wave propagation equally strong in all directions.

outer marker. Vertical signal cone, about 6 mi from end of runway; pilot sees purple light, hears 400 Hz audio tone Morse coded dash dash.

P code. Precision code—10 times faster than C/A code (not 10 times more accurate); military only.

parity. Addition of all binary bits to create even or odd numbers and thereby providing a method of checking the accuracy of data transmission.

port. The left side of the ship when facing forward.

PPS. P Code Positioning Service; term associated with GPS satellite signals; for military use.

preamble. A transmission that allows the receiver time to acquire the signal; provides aircraft with RF carrier acquisition and enables aircraft to receive time reference information.

PRF. Pulse Repetition Frequency; example: 9940 chain repeats its transmission of master pulse group every 99.4 μs.

Primary Surveillance Radar (PSR). Normal radar that relies on skin return.

Pseudo Station. An electronically generated station in an RNAV computer; see symbol. Also known as **Waypoint.**

Symbol for Pseudo Station ✧

Pseudo Random Noise (PRN). The phase of the GPS transmitted signal from the satellite is varied in a pattern that repeats over long periods of time. To a receiver that does not know the pattern, this appears to be noise.

quadrantal error. An ADF error caused by reradiation from the aircraft structure—wings, tail, props (prop modulation).

Radar. **RA**dio **D**etection **A**nd **R**anging.

radar mile. The distance that the RF energy propagates to the target and back to the transmitter. It takes 6.18 microseconds (μs) for RF energy to travel 1 NM. A radar mile is considered to be the distance there and back. The time duration of 1 radar mile = 12.36 μs.

radial. A magnetic direction away from a station.

Radio Magnetic Indicator. *See* **RMI.**

RAIM. Receiver Autonomous Integrity Monitoring.

Reference signal. 30 Hz Frequency Modulated.

relative bearing. The azimuth difference in degrees between the nose of the aircraft and a ground reference point (ground station).

reply. The transponder recognizes the interrogation and replies at a frequency of 1090 MHz.

reply light. Very small indication light on the front panel of the transponder that flashes during interrogation.

Resolution Advisory (RA). Computer response to intruder in immediate threat category; visual response—red indication on IVSI; aural response—"CLIMB, CLIMB, CLIMB."

resolver. A device with two stators wound 90° relative to each other, and one or two rotors.

RF. Radio frequency.

RF interference. Noise—man-made or environmental.

Rho Rho navigation. Distance/distance.

Rho Theta navigation. Distance/angle.

RMI. Radio Magnetic Indicator. Multipurpose indicator can display slaved Compass heading, ADF and VOR. Compass information displayed on the outside compass card. VOR information displayed on the #1 pointer indicating bearing of VOR selected station. ADF information displayed on the #2 pointer indicating bearing of ADF selected station.

RMI needles. Two needles will point at their respective selected stations on the ADF or VOR receivers; needles will always point at the station.

RNAV. *See* **Area Navigation.**

RVR. Runway Visual Range.

RX. Receiver.

Satellite Geometry. Relative location of the satellites that are visible to any given GPS receiver.

scanning beam. A high-frequency microwave signal.

scissors. A condition that occurs when the #1 ADF is selected to the destination station and the #2 ADF is selected to the departure station; when the needles of the RMI align, the plane is on the correct path between the two stations.

SDI. Source/destination identifier.

search mode. The warning flag showing that the DME is not operative.

Secondary Surveillance Radar (SSR). Allows for tracking of aircraft in control zones.

sector signals. Provides the aircraft with ground equipment identification, airborne antenna selection, OCI pulse, and "TO" pulse test.

selected course. The HSI setting for the desired radial along which to fly.

Selective Ability (SA). The American military's ability to manipulate the accuracy of the C/A code and the P code, thus directly influencing the system accuracy.

sense antenna. Vertical antenna (wire or whip).

servo loop. The electronic feedback system that controls the motor which powers the mechanical linkage connected to the load.

shore effect. An ADF error caused by a change in direction of a wave crossing from land to water or vice versa.

side lobe interrogation. The lessor power pulses from the SSR.

SIDS. Standard Instrument Departures.

sitting errors. Electronic errors due to the position of stations.

skin mapping. A surface analysis of the aircraft's magnetic fields to find where antennas interfere with each other, or where there are unusually large areas of magnetic influence that interfere with normal transmit/receiver operation.

slant range. Direct distance from the aircraft to the ground station.

slant range error. DME error based on line-of-sight distance to station and ground distance to station.

Space Segment. The orbiting satellite portion of the GPS.

SPI. Special Pulse Identification; this reply pulse is activated when the "ident" feature is selected.

SPS. Standard Positioning Service; term associated with GPS satellite signals; can be used by civilians.

Squawk Ident. A request by ATC for the pilot to select a code on the code selection switches and then press the Ident button, thus providing positive identification of the aircraft.

squitter. Random pulse generation from the DME ground station.

SSM. Sign Status Matrix from ARINC 429.

SSR. Secondary Surveillance Radar.

standard airway. Victor Airways are based on VOR radials, with a VOR station at the beginning and end of the airway.

starboard. The right side of the ship when facing forward.

STARS. Standard Terminal Arrival Routes.

STC. Supplemental Type Certificate; part of the certification process for aircraft.

STOL. Short Takeoff or Landing.

Suppression Bus. Interconnect between the transponder and DME such that only one transmits at any given time.

synchronous garble (sync garble). Overlapping of the replies from two aircraft from the same interrogator.

TACAN. A military station; Tactical Air Navigation station used to determine range and direction.

TCAS. Traffic Alert and Collision Avoidance System.

TO scan. On azimuth, moves from the far right to the far left; on elevation, moves to the ground.

track. The actual flight path of the aircraft.

TRACON (US). *T*erminal *R*adar *A*pproach and *Con*trol.

Traffic Advisory (TA). Computer response to intruder potential threat category; visual response—yellow indication on IVSI; aural response—"TRAFFIC, TRAFFIC, TRAFFIC . . ."

transponder. In general, a system that replies to an interrogation. Also see **ATC transponder.**

triangulate. By knowing the location and direction of two or more stations, it is possible to determine approximate aircraft position "fix."

TRSB. Time-referenced scanning beam.

TSO. Technical Standard Order—certification documentation.

TTS. Time to Station; the time displayed on the DME indicator for travel to the VORTAC station. This time is correct only if the aircraft is flying directly to the station.

twisted pair. Two single conductors used to transfer data twisted so as to reduce noise interference (common mode).

TX. Transmitter.

UHF. Ultra High Frequency, 300–3000 MHz.

USADOD. USA, Department of Defense.

User Segment. The receiver portion of the GPS.

Variable signal. 30 Hz Amplitude Modulated.

vertical effect. An ADF error caused by an unbalanced loop antenna circuit.

VFR. Visual Flight Rules.

VHF. Very High Frequency, 30–300 MHz.

VLF. Very Low Frequency, 9–14 kHz.

VOR. Very High Frequency Omni Range navigation system.

VOR bearing. Bearing information received from the VOR on board system.

VOR/DME. A VOR ground station colocated with a DME ground station.

VORTAC station. A VOR ground station colocated with a military TACAN station.

VSI. Vertical Speed Indicator.

walk around. A pilot's preflight check of the aircraft, usually done by a visual outside inspection.

waveguide. A transmission line for radio frequencies consisting of a hollow tube of fixed dimensions depending upon frequency.

Waypoint. A pseudo or fictitious station that can exist as a reference point in space.

WX. Weather radar.

Zone of Confusion. Inverted cone-shaped area above a VOR ground station that results in erratic VOR operation. *See* **Cone of Silence.**

INDEX